T0344473

Customizable and Adaptive Quantum Processors
Theory and Applications

Nadia Nedjah
Department of Electronics Engineering and Telecommunications
State University of Rio de Janeiro, Brazil

Luiza de Macedo Mourelle
Department of Systems Engineering and Computation
State University of Rio de Janeiro, Brazil

CRC Press
Taylor & Francis Group
Boca Raton London New York

CRC Press is an imprint of the
Taylor & Francis Group, an **informa** business

A SCIENCE PUBLISHERS BOOK

First edition published 2023
by CRC Press
6000 Broken Sound Parkway NW, Suite 300, Boca Raton, FL 33487-2742

and by CRC Press
4 Park Square, Milton Park, Abingdon, Oxon, OX14 4RN

CRC Press is an imprint of Taylor & Francis Group, LLC

Library of Congress Cataloging-in-Publication Data (applied for)

ISBN: 978-1-032-38229-6 (hbk)
ISBN: 978-1-032-38232-6 (pbk)
ISBN: 978-1-003-34409-4 (ebk)

DOI: 10.1201/9781003344094

Typeset in Times New Roman
by Radiant Productions

For all Students,
Regardless of Space and Time!

Preface

This book provides a representation of the state-of-the-art simulators and emulators that can be used for quantum computing. It especially reports a customizable and adaptive hardware design of a quantum processor that can be implemented in field programmable gate arrays to aid the traditional host processor to accelerate some computations and execute quantum algorithms. In the following, we give a brief description of the main contribution of each of the included chapters.

In the first chapter, entitled "Introduction to Quantum Computing", we introduce the motivation behind this book and we present and comment of recent related works in the literature about the quantum computing emulation by hardware and simulation by software. We also present the organization of the remaining chapters of the book.

In the second chapter, entitled "Quantum Concepts", we first introduce the main characteristics on the basis of quantum computing. After that we define and give illustrative examples for all the basic concepts that build the foundations of quantum computing. So, we define the indivisible piece of information that substitutes the traditional bit in the well-known classical computing paradigm. In the sequel, we explain and exemplify the parallelism behind any quantum-related computation. Subsequently, we describe the process of quantum information sampling also know as measurement of the quantum state of a given piece of information. Last but not the least, we explain the process of quantum bit entanglement, collapsed state as well as the process of information cloning in contrast to information copying in the traditional computing paradigm.

In the third chapter, entitled "Quantum Operators", we first introduce the concept of quantum operators. After that, we define the main tensor product operation, which is behind almost all quantum information manipulation. Subsequently, we present existing unitary, binary and ternary quantum operators as well as the computation implied by every one of the presented operators. We also explain that the tensor product is not an end calculation in a quantum operation, it has significant importance due to the overall number of multiplications with complex numbers that it requires. We prove that this number increases exponentially with the number of quantum bits involved in the quantum state. We show that if not handled efficiently, it would require a lot of memory space, which makes the design emulating a quantum processor quite expensive to produce.

In the fourth chapter, entitled "Quantum Processor Macro-architecture", we present the overall hardware architecture of a quantum processor. First of all, we show its macro-architecture, explaining the role of each of its structuring components. Emulation of quantum processing primarily requires an efficient way to store and retrieve quantum information. The quantum processor employs different purpose memories. There is the quantum state memory, which stores information about the possible states of the quantum bits; the quantum operator memory, which stores the quantum operators in their basic form; the scratch memory, which allows storing intermediate results regarding the computation of higher-dimensionality operators form the basic ones, when these are required. For each of these memories, we describe the format used to organize the required information and show simulation results about their usage both in read and write cycles.

In the fifth chapter, entitled "Calculation Unit Micro-architecture", we first give an overall view regarding the function that must be executed by the calculation unit. We present and explain the underlying micro-architecture of this functional unit. The task of this unit is to allow an efficient computation of tensor and matrix products, which are required by any quantum information manipulation made by the processor. It also allows the computation of complex number summations. For this purpose, the unit implements efficient parallel complex number multiplications, which form the foundations of tensor and matrix products. After that, we explain and give illustrative an example on how the calculation unit proceeds and controls the computation of the tensor product of quantum bits and registers. Then, we do the same for the tensor products of operators, required to obtain an operator of high dimensionality from the basic one pre-stored in the quantum operator memory. Last but not least, we explain how the calculation unit proceeds to provide a required matrix product.

In the sixth chapter, entitled "Control Unit Micro-architecture", we present the control unit of the processor. We first describe the micro-architecture of this unit and define the main control components required to implement the execution of the quantum micro-instructions. We also define the format of such micro-instructions. We then explain and exemplify the flow within the control unit during the execution of a given quantum instruction. After that, we do the same regarding the dynamics during the execution of the micro-instructions that interpret the quantum instructions. Subsequently, we present the tensor product controller, which controls the three main operations: tensor product of quantum bits; matrix product of quantum bits; tensor product of operators. Along with the calculation unit, this unit forms the heart of the quantum processor. It performs the expected control through micro-instructions.

In the seventh chapter, entitled "Quantum State Measurement", we propose a possible simple yet efficient implementation of the quantum state sampling unit. First, we present the idea behind the solution. Then, we describe and explain the micro-architecture of the measurement unit. It performs the measurement of the quantum state using the roulette selection algorithm and pseudo-random number generator.

In the eighth chapter, entitled "Quantum Instruction Simulation", we give proof of the correct operation of the proposed processor. For this purpose, we present a simulation about every aspect of the quantum processor while executing a quantum instruction. We recall the main elements of a quantum instruction and their usage during the execution process. Then, we illustrate the execution of a quantum operation that manipulates a single quantum bit, showing the time diagram during its simulation. After that, we illustrate the execution of an operation on non-entangled quantum bits, showing the time diagram during its simulation. It is followed by the execution simulation of a quantum operation on entangled quantum bits. Subsequently, we illustrate the building of a high-dimensionality quantum operator from the predefined basic operator, showing the time diagram during its simulation. This is shown for two and three quantum bits. Then we, do the same for the tensor product of quantum bits, showing the time diagram during its simulation. Last but not the least, we illustrate it via the simulation of the matrix product of entangled quantum bits and the measurement of the quantum state of the obtained result. This concludes the proof via simulation that the proposed customizable quantum processor is functional and can be used to run quantum algorithms on general-purpose computers, allowing the evaluation of their effectiveness and performance.

In the ninth chapter, entitled "Book Final Remarks", we draw some final considerations about the book and point out some directions on how to improve the current state of the art of quantum computing simulators/emulators.

The authors of this book are extremely grateful to the students for their valuable contributions. The authors would also like to thank the editorial team that helped formatting this work into a good book. Finally, we sincerely hope that the reader will share our excitement with this book on the emulation of quantum computing and will find it useful.

Nadia Nedjah, Ph.D. & Luiza de Macedo Mourelle, Ph.D.
Engineering Faculty
State University of Rio de Janeiro
Rio de Janeiro, RJ, Brazil

Contents

SECTION I: THEORY

SECTION II: APPLICATIONS

List of Figures

List of Tables

List of Algorithms

Symbol Description

$|0\rangle$	A q-bit in the so-called collapsed state 0.	noq_o	For an available basic operator o, the number of q-bits it operates on.
$|1\rangle$	A q-bit in the so-called collapsed state 1.	nop	The total number of basic operators.
$\alpha, \beta, \gamma, \omega$	Symbols used to reference amplitudes of q-bits.	aop	The number of required bits to address operator memory.
mq	The maximum number of q-bits that the quantum processor can operate on simultaneously.	FPGA	Field programable gate array.
		QCA	Quantum-dot cellular automata.
oq	The number of q-bits that the current operator handles.	QCS	Quantum computer simulator.
		QCL	Quantum computation language.
nq	The maximum number of q-bits that the co-processor can have in its state.	QML	Quantum machine learning.
		I	The identity unitary operator.
		H	The Hadamard unitary operator.

S	The S unitary operator.	ROM	Read-only memory.
T	The T unitary operator.	MDATA	The bidirectional measurement bus data.
X	The X unitary operator.	SDATA	The bidirectional scratch data bus.
Y	The Y unitary operator.	ODATA	The unidirectional operator data bus.
Z	The Z unitary operator.	QDATA	The bidirectional quantum bus data.
CNOT	The controlled NOT binary operator.	OADDR	The operator memory address bus.
CSWAP	The controlled swap ternary operator.	QADDR	The quantum state memory address bus.
NOT	The NOT binary operator.	RANDOM	The unidirectional random number bus.
SWAP	The swap binary operator.	SADDR	The scratch memory address bus.
TOFOLLI	Tofolli's ternary operator.	UCA	The calculation unit.
MSC	The scratch memory.	UCO	The control unit.
MOP	The operator memory.	UMS	The measurement unit.
MQB	The quantum state bit memory.	EvenRe	Even real part of a complex number.
MQC	The quantum state control memory.	EvenIm	Even imaginary part of a complex number.
MQS	The quantum state memory.	OddRe	Odd real part of a complex number.
COPROC	The quantum processor	OddIm	Odd imaginary part of a complex number.
PROC	The host processor		
RAM	Random access memory.	FPU	Floating-point unit.

MIR	Micro-instruction register.	LFSR	Linear feedback shift register.
MPR	Micro-program memory.	PRNG	The pseudorandom number generator.
IEEE754	Floating point IEEE standard.	RNG	The random number generator.

THEORY

I

Chapter 1

Introduction to Quantum Computing

In this chapter, we introduce the motivation behind this book and we present and comment on recent related works in the literature about the quantum computing emulation by hardware and simulation by software. We also present the organization of the remaining chapters of the book.

1.1 Introduction

The conventional electronic computer is a machine that performs logical and arithmetic operations sequentially and/or in parallel on data coded according to binary logic. This Cartesian model for the elementary information unit, which is the bit, is very convenient for processing data in an electronic circuit, as the two possible states can be represented in a very well controlled and safe manner by digital circuits.

According to the updated version of Moore's Law, originally coined by Roger Moore in 1965, the processing capacity of computers doubles over every twenty-four month period [3]. Accelerating the execution of instructions can be achieved in several ways, such as increasing the clock frequency, using smart peripherals to reduce demand on the main processor, using multiple processors and/or more than one processor, more advanced processor architectures (Harvard × Von Neumann), processors with simpler instructions, typically RISC, use of faster memories among many other strategies for computer system acceleration. In the last 50 years, Moore's law has been fulfilled, but if a new technology does not appear, the validity of this Law is threatened due to the approximation of hard physical limits [3]. The increase in the performance of processors by increasing the number and density of transistors is approaching the feasible limit for current technology. This is not only due to the complexity of developing and producing processors with nano-metric precision on an industrial scale, with implications regarding energy consumption, heat dissipation and precision during the manufacturing process of tiny devices, but also by approaching a quantum context. In other words, the small dimensions of a processor's components tend to depart from the predictable and well-known rules of classical physics. In [44] the availability of 5nm technology is mentioned and 3nm is already cited as available for production in late 2022 [22].

Quantum Computing has been the target of several researches due to its potential for increasing the speed of processing through the use of algorithms with intrinsic parallelism, pointing out to the possibility of polynomial time solutions for *NP*-Complete problems [6, 30, 1]. On the other hand, the technology of quantum computers still seeks ways to control electrons, which has already been achieved with only a few of them and only for a short time. At the time of the conclusion of this work, the largest number of entangled q-bits reached under special conditions is twenty [15]. Investments by companies have already resulted in commercial devices, as is the case with the In 2015, with the D-Wave company advertising a quantum computer with 1,000 q-bits [10], but not entirely entangled. The primary focus of the invented device is to execute algorithms for adiabatic computing [18, 13, 2, 8]. In 2019, the same company announced their next-generation Pegasus quantum processor chip, with 15 connections per qubit instead of 6, and that the next-generation system would use the Pegasus chip, which would include more than 5000 qubits, and that it would be available in mid 2020.

At the same time, the technology for chip production is reaching the limit of miniaturization, approaching the frontier that separates the world of classical abstraction from quantum physics. As long as there are no commercial quantum processors, quantum programming is being tested through emulators and libraries of routines for quantum operations, of which examples such as QCL (Quantum Computation Language) [36], QCS (Quantum Computer Simulator), QuaSi, Fraunhofer Quantum Computing Simulator, QuCalc, QDensity, OpenQuacs, QML, JaQuzzi, Senko's Quantum Computer, Shornuf, SimQ-bit and QHaskell [3, 50, 28]. We highlight the Jülicher Supicercomputer Jugene simuliert, which is a simulator for 42 q-bits, developed by the research group Jülich and a Computational Physics group at the University of Groningen, in the Netherlands, at the moment still undefeated in quantity terms of entangled q-bits [29]. Note that R. Feynman was the first to suggest that we ought to simulate physics using computers, including classical physics and quantum mechanics [14], which later was emphasized in [12, 54].

Simulators, which are software products, imply greater processing time due to the sequential execution, eventually accelerated by the use of multiple processors, and the sharing of resources with other processes, than if it were run by dedicated hardware, which would deliver the results of quantum operations in shorter time, thanks to the faster speed and parallelism in the operation execution [33, 48]. Such hardware could take the form of a specialized processor, which here is considered as a quantum co-processor, which would be part of a larger computer dotted by a conventional processor. This work presents the architecture of such a quantum co-processor that can be implemented on hardware and embedded into our personal computers to accelerate the resolution of hard complex problems. As far as the authors are concerned, this work is a pioneer as it is the first and only attempt at a hardware implementation of a quantum co-processor that executes quantum operation as user directives inserted within a traditional program.

1.2   Current State

Several research works in the literature have invested in the simulation of quantum operations both in software and hardware. Some of these works implement libraries, quantitative circuits using programmable or reconfigurable devices, initiatives not limited to the use of hardware description languages, including works on designing modeling tools.

1.2.1   Quantum Hardware

In [21], a quantum circuit model to describe the main quantum algorithms and the corresponding analogies with the digital circuit model is proposed. They introduce an emulator of quantum algorithms in FPGA, concentrating their effort on new techniques for modeling quantum circuits, including the treatment of q-bit entanglement, probabilistic computation and critical precision issues.

In [45], the authors analyze the logical efficiency of quantum circuits that perform generic quantum computations and initialization of quantum registers.

In [55], the authors propose an approach to synthesize quantum circuits from non-permutative quantum gates, such as controlled square root of not quantum gate, commonly termed controlled-V. Group theory is used in order to transform the synthesis problem to a multiple-valued optimization.

In [19], the authors explore the nano-technology of quantum-dot cellular automata (QCA) to propose a new reconfigurable device (FPGA) with efficient, symmetric and reliable programmable switch matrix interconnection elements. The results demonstrate high efficiency of the proposed designs in QCA-based FPGA routing.

In [48], the authors discuss the challenges faced during the design of a scalable electronic interface for quantum processors, as well as the requirements required depending on different q-bit existing technologies.

1.2.2 Quantum Software

In [38], the author investigates, how classical concepts like hardware abstraction, structured programs and data types, memory management, control flow can be exploited in quantum computing. He proposes a quantum computing language, called QCL. A QCL interpreter is made available to run quantum programs. It also includes non-quantum related instructions, such as irreversible functions, local variables and conditional branching. Using the provided interpreter, one can experiment on how non-classical features like the reversibility of unitary transformations or the non-observability of quantum states can be accounted for in a procedural programming language.

In [20], a quantum simulator is proposed. It is intended for users not having a deep understanding of quantum mechanics. The simulator is based on circuit model of quantum computation in which quantum gates act on quantum registers which comprise a number of quantum bits.

In [11], a massively parallel quantum computer simulator is proposed. It uses a software component with a portable feature to simulate universal quantum computers using parallel computers. The simulation in software encompasses several quantum algorithms in different computer architectures. The simulator outputs are the matrices that represent the quantum register state at each quantum computation step and details that show the measurement probability of each used quantum register. The well-known Deutsch's algorithm and the quantum Fourier transform are presented using the proposed simulator.

In [27], the authors optimize the execution library of the designed visual programming environment for the quantum geometric machine model. The model uses recursion of mathematical functions to dynamically generate values that define quantum transformations, obtaining considerable decrease in memory consumption.

In [33], a general direct simulator for one-way quantum computation (1WQC) model, called OWQS, is proposed. Some techniques such as qubit elimination, pattern reordering and implicit simulation of actions are used to considerably reduce the time and memory needed for the simulations. Moreover, the measurement patterns with a generalized flow without calculating the measurement probabilities is employed. Experimental results validate the effectiveness and efficiency of the proposed model for quantum computing simulation.

In [52], the authors present some useful extensions for the programming language for Synthesis of Reversible Circuits (SyReC), which allows for the specification and automatic synthesis of reversible circuits. They also propose algorithms that optimize the resulting circuits with respect to different objectives, such as time delay and circuit cost.

In [16], the authors propose a control architecture that allows fault-tolerant quantum computing based on the rotated planar surface code with logical operations. It exhibits a two-level address mechanism that yields an adequate compilation model for a large number of quantum bits. It also has an architectural support for quantum error correction at runtime, significantly reducing the size of the quantum program and improving its scalability.

1.3   Book Organization

The remaining paper is organized into seven chapters. The book covers all the required definitions to understand the proposed design of the hardware architecture of the quantum processor.

First, in Chapter 2, we introduce and define quantum computing, the model of the handled data and the main quantum operations, thus covering a complete definition of quantum computation. These include the definition of the basic unit of information, which is the quantum bit, the composition of quantum bits to form a quantum register. We explain the intrinsic quantum parallelism and the observation of quantum bits for measurement.

After that, in Chapter 3, we define the quantum operators. These are classified into three categories: unitary, binary and tertiary operators, depending on how many quantum bits they require to operate.

Then, in Chapter 4, we describe the macro-architecture of the proposed quantum processor and its interaction with the conventional host wherein it is embedded. For each unit of the macro-architecture, we give a brief explanation of its role within the design, including the role of each used data, address and control bus. Also, we describe the functionality of the three main memories used, which are the quantum state memory, the quantum operator memory and the scratch memory.

Subsequently, in Chapter 5, we go through the micro architecture of the quantum processor unit that is dedicated to calculation. For this purpose, we describe the micro-architectures designed for the units that efficiently compute the tensor product of quantum bits and that of operators. The latter allows the construction of complex operators form basic ones. We also, present the micro-architecture for the unit responsible for the computation of the matrix product, which produces the final result of an operation on specified quantum bits.

After that, in Chapter 6, we show and describe the operation of the control unit of the quantum processor. So, we explain the instruction flow within the quantum processor. After that, we describe the micro-instruction flow as executed to interpret a given quantum instruction. Then, we present the micro-architecture of the used controllers, which are aimed at controlling the tensor and the matrix product computations.

In the sequel, in Chapter 7, we define quantum state measurement and propose an efficient micro-architecture for the unit that is responsible for the state observation of the quantum bits, thus allowing for quantum bit measurements.

Later on, in Chapter 8, we present, analyze and discuss the simulation results regarding the execution of quantum instructions within the quantum processor. We show step by step, via simulation, how quantum instructions would be executed in the proposed quantum processor.

Last but not least, in Chapter 9, we drawn some conclusions to point out some promising directions for future improvements of the proposed quantum processor.

1.4   Chapter Considerations

In this chapter, we give a brief overview on the state of the art in the published literature regarding the usage of quantum algorithms while there are no real quantum computers. This can be done by emulation using quantum processors implemented on hardware or a software tool that simulates the execution of quantum operations.

In the next chapter, we define the basic concepts of quantum computing, such as quantum bit, quantum register, quantum parallelism, quantum state measurement, quantum state entanglement and quantum state cloning.

Chapter 2

Quantum Concepts

In this chapter, we first introduce the main characteristics at the basis of quantum computing. After that we define and give illustrative examples for all the basic concepts that build the foundations of quantum computing. So, we define the indivisible piece information that substitutes the traditional bit in the well-known classical computing paradigm. In the sequel, we explain and exemplify the parallelism behind any quantum-related computation. Subsequently, we describe the process of quantum information sampling also know as measurement of the quantum state of a given piece of information. Last but not the least, we explain the process of quantum bit entanglement, collapsed state as well as the process of information cloning in contrast to information copying in the traditional computing paradigm.

2.1 Introduction

At a macro level, a traditional computer can be generally described as a machine that reads data, performs calculations and generates data output, which is binary encoded data, commonly represented by 0s and 1s that are values associated with low and high voltage levels, respectively. On the other hand, classical computing has the bit as its basic information unit, which can assume one among two states at a given time. Until the storage of a different state (or binary value) is commanded, the bit will maintain its previous state *ad infinitum* as long as the hardware remains energized [51]. Generally speaking, the energy present at the inputs does not equal the energy at the outputs, and the difference is converted into heat. A bit is a data repository independent of other bits, so that changing its content does not affect that of the other bits and *vice versa*. A bit can be read infinitely without changing its status. So, the input data is processed by logical or arithmetic units that result in an output, which is equivalent to stating that such units correspond to functions in a mathematical sense. These units perform more or less complex logical operations, built by appropriate elementary gate circuits such as NOT, AND and OR [51]. Some aspects of classical computing are worth highlighting when considering their counterparts in quantum computing. In classical computing, the sequential characteristic of executing instructions stored in memory is also intrinsic to the aforementioned computational model. Even if parallel processors, peripherals or subsystems are used with their own processors, the sequential characteristic of executing the program stored in memory remains. With the exception of the NOT gate, the gates AND, OR and derivatives do not allow knowledge of states present at their input pins, taking only the state present at the output pins. For example, an OR gate with two inputs, when its output exhibits value of 1, it cannot be said whether only one of the inputs or both are at logic level 1. For this reason, the classic gates are said to be *irreversible*.

Quantum mechanics is the part of physics that studies and describes the behavior of particles at the atomic and subatomic levels, complementing classical physics, which deals with concrete reality. This is based on Newtonian Mechanics and Electro-magnetism, based on wave and matrix mechanics, later unified by the description of Paul Dirac [35]. A quantum computer performs calculations using the properties of quantum mechanics, which is a probabilistic context, unlike the deterministic condition assumed for classical computing, although it can also perform the same operations. In quantum computing, there is an intrinsic parallelism, since the quantum state is a superposition of basic states [46]. Data repositories can establish a relationship with mutual interference, known as *entanglement* [35].

2.2   Quantum Bit

The basic information unit in a quantum computer is the *quantum bit* or simply put the *q-bit*[1], capable of storing not only states 0 or 1 as in the case of a classic bit, but both, each with its own magnitude or *amplitude* [39, 37, 34]. Such states are said to be in *superposition*. In other words, a q-bit can take on infinite values, including the boundary states 0 and 1, with probabilistic rather than deterministic characteristics. A q-bit can be viewed as a vector with an orthonormal base[2] of 2 dimensions, able to represent infinite values by means of linear combination in the field of complex numbers, marked out by the base vectors [51, 34, 5], as denoted in Equation 2.1:

$$|0\rangle = \begin{bmatrix} 1 \\ 0 \end{bmatrix} \text{ and } |1\rangle = \begin{bmatrix} 0 \\ 1 \end{bmatrix}. \tag{2.1}$$

So, q-bit $|v\rangle$ can be represented interchangeably by either of the forms shown by Equation 2.2:

$$|v\rangle = \begin{bmatrix} \alpha \\ \beta \end{bmatrix} \text{ or } |v\rangle = \alpha|0\rangle + \beta|1\rangle, \tag{2.2}$$

wherein α and β are complex numbers that denote the magnitude of each base vector. Thus, it is allowed to say that a q-bit can simultaneously represent states 0 and 1, but with their own coefficients for each one of the states. The amplitude squared gives the probability (in the closed interval $[0, 1]$) that the q-bit is in that respective state [51], according to the Born's rule [25, 24]. The sum of the squares of the amplitudes of each possible state is always 1, according to Equation 2.3:

$$|\alpha|^2 + |\beta|^2 = 1, \tag{2.3}$$

wherein the vecter norm is preserved [49].

Thus, the state of a quantum system can now be represented by a vector that is a linear combination of the vectors of the base, with the base dimension equal to 2^n, where n is the amount of q-bits in the system. Therefore, a vector representing the state of the quantum

---

[1] Note that *qbit, qubit qubit* are other abbreviations found in the literature.

[2] Orthonormal base is a set of linearly independent vectors, orthogonal two by two, each with norm 1, and that generate a vector space.

system does not only hold as many states as the number of q-bits, but 2^n states. The commonly adopted way to represent a q-bit is the Dirac notation, also known as *braket*, the English name for the characters "⟨" and "⟩". The notation includes the character "|" between brackets, termed *bra* and the q-bit between the "⟨" or the "|" and "⟩" termed it ket [47, 49].

Kets are used to represent a possible state of the quantum system composed of one or more q-bits. A q-bit in the so-called collapsed state, that is, with 100% probability of being in state 0 or 1, is represented as $|0\rangle$ or $|1\rangle$, respectively. A state composed of more than one q-bit is written with as many digits from the set $\{0, 1\}$ as those of q-bits, whose order grows from right to left, as in a binary representation [42, 49]. This groupings of q-bits is called *quantum register* [51, 37, 5]. For instance, we have:

- Two q-bits: $|01\rangle$ means that the less significant q-bit is in state 1 and the other in state 0;

- Three q-bits: $|110\rangle$ means that the less significant q-bit is in state 0 and the other two are in state 1.

An alternative way of representing a set of q-bits uses the decimal base instead of the binary one. For instance, the example above with the set of q-bits $|110\rangle$ could be represented as $vert6\rangle$. The basis for representing a set of n q-bits has 2^n vectors or states. So a set of n q-bits can be represented as in Equation 2.4:

$$|v\rangle = \alpha|0\rangle + \beta|1\rangle + \gamma|2\rangle + \cdots + \omega|2^n - 1\rangle, \qquad (2.4)$$

wherein a unit vector is the sum of the squares of the amplitude modules which should result in 1, as shown in Equation 2.5:

$$|\alpha|^2 + |\beta|^2 + |\gamma|^2 + \cdots + |\omega|^2 = 1. \qquad (2.5)$$

A spherical coordinate system, as shown in Figure 2.1, serves as the basis for the graphical representation of the state of a q-bit, known as Bloch's Sphere [9]. Such a polar representation in $\mathbb{C}^3$ makes it possible to visualize the quantum state and the infinite possibilities that could result with a probability p of a measurement resulting in $|0\rangle$ or probability $1 - p$ of resulting $|1\rangle$. The projections on the axes

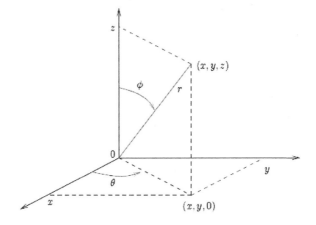

Figure 2.1: Spherical coordinate system.

x, y and z are defined by Equation 2.6:

$$\begin{bmatrix} x \\ y \\ z \end{bmatrix} = \begin{bmatrix} cos\phi.sin\theta \\ sin\phi.sin\theta \\ cos\theta \end{bmatrix}, \qquad (2.6)$$

wherein $0 \leq \theta \leq \pi$ and $0 \leq \phi \leq 2 \times \pi$. So, The basis for the representation in $\mathbb{C}^3$ is defined as in Equation 2.7 [41]:

$$|0\rangle = \begin{bmatrix} 0 \\ 0 \\ 1 \end{bmatrix} \quad \text{and} \quad |1\rangle = \begin{bmatrix} 0 \\ 0 \\ -1 \end{bmatrix}. \qquad (2.7)$$

Thus, considering the spherical representation of coordinates, it can be observed that the collapsed state $|0\rangle$ results in a vector pointing to north and the collapsed state $|1\rangle$ to south [41]. As the vector is applied in the center of a sphere, its norm (length) remains constant as 1, which is consistent with the sum of the squares of the amplitudes of the two possible states (base vectors) [41].

Let us consider $\alpha = cos(\theta/2)$ and $\beta = \varepsilon^{i\phi} sen(\theta/2)$ [41]. The representation of the state of a q-bit can also be viewed in the trigonometric form, shown in Equation 2.8:

$$|v\rangle = cos(\theta/2)|0\rangle + \varepsilon^{i\phi} sin(\theta/2)|1\rangle. \qquad (2.8)$$

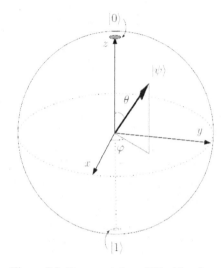

Figure 2.2: Representation of Bloch's sphere.

Table 2.1 shows the direction and orientation of the representative vector of the q-bit for some states [34, 9].

Table 2.1: Examples of direction and orientation of the representative vector of the q-bit for some states.

θ	ϕ	$|v\rangle$	Observation
0	0	$|0\rangle$	North pole
π	0	$|1\rangle$	South pole
$\frac{\pi}{2}$	0	$\frac{|0\rangle+|1\rangle}{\sqrt{2}}$	Equator on axis x
$\frac{\pi}{2}$	$\frac{\pi}{2}$	$\frac{|0\rangle+i|1\rangle}{\sqrt{2}}$	Equator on axis y

2.3 Quantum Parallelism

A linear combination whose terms are collapsed states (base vectors) multiplied by the associated amplitude forms the quantum state of the machine [4]. An operation on a vector is equal to the sum of the operation on each term of the linear combination. So, an intrinsic parallelism is configured, and for any given operator, say T, we have Equation 2.9:

$$
\begin{aligned}
T|v\rangle &= T(\alpha|0\rangle + \beta|1\rangle + \gamma|2\rangle + \cdots + \omega|2^n - 1\rangle) \\
&= \alpha T|0\rangle + \beta T|1\rangle + \gamma T|2\rangle + \cdots + \omega T|2^n - 1\rangle.
\end{aligned}
\tag{2.9}
$$

A quantum logic gate allows the evaluation of more than one state simultaneously [37], as illustrated in the following example. Let $|\phi\rangle$ be a quantum register[3] for 2 quantum q-bits, as defined in Equation 2.10:

$$|\phi\rangle = |a\rangle|b\rangle. \tag{2.10}$$

For unitary operator T, we have Equation 2.11:

$$T(|\phi\rangle) = T(|a\rangle, |b\rangle) = |a\rangle|b \oplus f(a)\rangle, \tag{2.11}$$

wherein, $\oplus$ represents a *modulo-2* sum, $a, b \in \{0, 1\}$, $f(a) : \{0, 1\} \to \{0, 1\}$ is the underlying function for operator T. Assuming that $|a\rangle$ is in a superimposed state such as in Equation 2.12:

$$|a\rangle = \frac{|0\rangle + |1\rangle}{\sqrt{2}} \quad \text{and} \quad |b\rangle = |0\rangle. \tag{2.12}$$

So we would have Equation 2.11 applied as in Equation 2.13:

$$T(|\phi\rangle) = |a\rangle|0 \oplus f(a)\rangle = |a\rangle|f(a)\rangle = \frac{|0\rangle|f(0)\rangle + |1\rangle|f(1)\rangle}{\sqrt{2}}. \tag{2.13}$$

Hence, in this example, we illustrate the possibility in which q-bit a is in a superimposed state, that is, two fundamental states ($|0\rangle$ e $|1\rangle$) present simultaneously, a quantum gate would act on all states simultaneously, which characterizes the intrinsic quantum parallelism.

As an illustration, we consider the example of a q-bit a in a superposition state, *i.e.*, the two fundamental states ($|0\rangle$ and $|1\rangle$) present simultaneously, so that a quantum gate would act on all states simultaneously. Figure 2.3 shows the classic version using classical gates.

---

[3] Quantum register is a set of q-bits, which can use the matrix notation of a vector column, with 2^q rows, where q is the number of hypothetical q-bits.

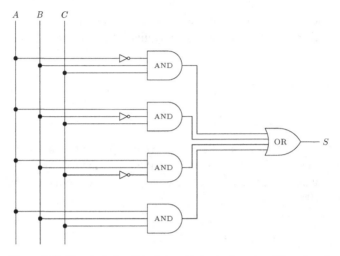

Figure 2.3: Classical circuit to determine majority vote of three inputs.

As classical gates are used, the evaluation of the eight possible states (considering the existence of three independent inputs: A, B and C), could only occur at sequential times, requiring more than one iteration of the algorithm or the performance of more processors [28]. This circuit has the same logical output state as that of most of its input signals A, B and C. The truth table of this circuit is shown in Table 2.2. Figure 2.4 shows the quantum version of the voting circuit of Figure 2.3. It is a typical example of the application of the simultaneous performance in a quantum majority voting circuit [32].

Table 2.2: Truth table of the circuit of Figure 2.3.

A	B	C	S
0	0	0	0
0	0	1	0
0	1	0	0
0	1	1	1
1	0	0	0
1	0	1	1
1	1	0	1
1	1	1	1

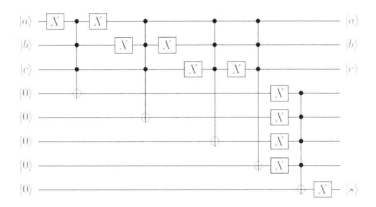

Figure 2.4: Quantum circuit to determine the majority vote of three inputs.

In Figure 2.4, the boxed Xs represent quantum operator NOT gates. It is an equivalent of the classic NOT gate. In each of the control lines *A*, *B* and *C* there is a second quantum operator NOT whenever it is needed to restore the original state of the control q-bit. The connections represented by black dots refer to the control q-bits and the connection represented by ⊕ refers to target q-bits of a generalized Toffoli port, which is explained in detail in Chapter 3. For now, it is enough to understand that a Toffoli gate operates as an inverting quantum port, which only acts on the target q-bit when the control q-bit is in state 1. The input signals *A*, *B* and *C* receive the q-bits that represent the votes and act as controls for the Toffoli gates. The five inputs receive a quantum data collapsed in $|0\rangle$ and are used as targets for the Toffoli gates. Each of the first three Toffoli gates, counted from left to right, deals with a case of a double vote $|1\rangle$, respectively *B* and *C*, *A* and *C* and *A* and *B*. The fourth Toffoli gate solves the case where all control inputs are in state $|1\rangle$. The fifth and final Toffoli gate inverts the state of the last line $|0\rangle$ according to the given vote. This result is inverted in a quantum way and delivered at the output $|s\rangle$.

Using as a specific example of configuration as the case in which *B* and *C* are in state $|1\rangle$ and *A* in state $|0\rangle$, the first Toffoli gtae, having value $|1\rangle$ in all its control inputs, will invert the state $|0\rangle$ of the provided target q-bit, changing it to $|1\rangle$. Gate X on the same row will invert the state to $|0\rangle$, which will prevent the last Toffoli port in the circuit from inverting the state of its target q-bit, remaining at state $|0\rangle$. Finally, the gate X in the list row of the same target q-bit will take care of applying state $|0\rangle$ to the output $|0\rangle$. Similarly, the analysis is applicable to other possibilities in which 2 or more control inputs are in the state $|1\rangle$.

2.4   Quantum State Measurement

Measuring the state of a q-bit results in changing the probabilistic condition to assume a collapsed state, which is one of the pure binary states, like in the case of a classic bit. Therefore, reading the status of this quantum information unit simply results in a value of 0 or 1 [47, 42], consistent with the probabilities associated with each state, unlike the classic bit.

After a measurement, a set of n q-bits ($n \geq 1$) can assume one of 2^n possible states, with a probability associated with each state, given by the squared amplitude relative to the respective state [51]. As the q-bit measurement, or q-bit observation, interferes with the quantum system, the observer, which is also a quantum system, the observed q-bit loses its probabilistic condition and assumes one of the collapsed states $|0\rangle$ or $|1\rangle$.

In a real quantum system, the probabilities associated with each q-bit cannot be known at runtime, as this would mean observing the q-bit and, therefore, ceasing the quantum condition. The programmer of a quantum algorithm is responsible for controlling the probabilities, using techniques that allow for yielding the expected result in this non-Cartesian context. In a simulator or emulator for quantum computing, the knowledge of these probabilities is provided to the programmer, without resulting in the destruction of the quantum characteristic.

2.5   Entanglement and Cloning

A q-bit can relate to one or more other q-bits in order to establish mutual interference, called *entanglement*, and not merely a water-tight group of q-bits [41]. Although entangled q-bits do not cease to exist individually, entanglement establishes a grouping characteristic that is of interest in Quantum Computing, such as super-dense coding and teleportation [34, 43]. The reading of a tangled q-bit results in the collapse of all q-bits of the group, leading them to the solidarity assumption of a value of 0 or 1 [35]. The tangled condition of 2 or more q-bits is not a factoring property. This means that there are no individual states of q-bits that after any state manipulation can result in a grouping state. In this way, tangled q-bits are operated by appropriate quantum gates to act on the set of q-bits [24].

Unlike classical computing, q-bit cannot be copied with impunity, as this would represent a measurement and, therefore, would result in a q-bit condition change. In [53], the Non-Cloning Theorem can be explained as follows: suppose a machine with an entry for 2 q-bits, the first being a unknown state $|\phi\rangle$ and the other an "inert" state $|v\rangle$, like a sheet of virgin paper in the feed tray of a photocopy machine. The initial state of this machine is $|\phi\rangle \otimes |v\rangle$, wherein $\otimes$ represents the tensor product operation. One would like to discover a unitary operator T such that, the state at the output of the machine is $|\phi\rangle \otimes |\phi\rangle$. This means that the second q-bit takes the same state as the first, without changing it. The hypothetical operator T would have to be able to reproduce in $|v\rangle$ the state of $|\phi\rangle$, independently of its state. The usual internal product[4] between $|\phi\rangle$ and $|v\rangle$, which would result in $\langle\phi|v\rangle^2$, would only be useful when it resulted into either state 0 or 1, respectively when either the states are orthogonal or equal. For any other result, operator T operator would fail [53].

2.6 Chapter Considerations

In the next chapter, we explain the basic concepts of quantum computing. The q-bit universal definition is thoroughly explained as well the composition of q-bits to compose quantum registers. We also explain the quantum parallelism that occurs when quantum gates are executed on q-bits. Moreover, we explain the process that occurs when a quantum state is measured to obtain the collapsed state of a quan-

---

[4]Let $x = (x_1, x_2, x_3, \ldots, x_n)$ and $y = (y_1, y_2, y_3, \ldots, y_n)$ be two vectors in $\mathbb{R}^n$. Usual internal product is the operation defined by $\langle x, y \rangle = x_1 y_1 + x_2 y_2 + x_3 y_3 + \cdots + x_n y_n$ and meets the properties of positivity, additivity, homogeneity and symmetry, whose descriptions falls out of the scope of this work. The usual Internal Product is also known as *canonical internal product* or *Euclidean internal product* [47].

tum bit or register. Finally, we explain in detail the notion of quantum state entanglement as well as quantum state cloning.

In the next chapter, we define quantum operators and their compositions as well as the involved mathematical operations therein.

Chapter 3

Quantum Operators

In this chapter, we first introduce the concept of quantum operators. After that, we define the main operation of tensor product, which is behind almost all quantum information manipulation. Subsequently, we present existing unitary, binary and ternary quantum operators as well as the computation implied by all the presented operators.

3.1 Introduction

Quantum operators allow the execution of operations on one or more q-bits. They have an equal number of inputs and outputs, maintaining the equivalence between the energy of the inputs and that of the outputs. Therefore, there should be no heat dissipation. These operators allow knowledge of the conditions of the input data, as this information is preserved. When the operators are reversible, it is possible to return the quantum system to its previous state, *i.e.*, before applying the quantum operator in question [51, 42].

Quantum operators have matrix representation as unitary operators [51, 35]. Recall that, basic quantum operators have dimensions 2×2 and process the state of one q-bit, with the exception of the controlled NOT (CNOT) operator, whose size is 4×4 as it acts basically on two q-bits, and Toffoli's gate, whose size is $2^i \times 2^i$, where i is the number of control inputs plus 1, Swap and controlled Swap (CSwap) operators, as defined later in this section. Quantum operators can act on two or more q-bits simultaneously. In this case, the operator is built using tensor products, as explained in the previous section. In this way, a quantum operator, with the exception of CNOT or Toffoli's gate, can be constructed from tensor products between basic quantum operators, which imposes the state reversibility characteristic. That is, from a certain quantum state, one can return to the previous state by applying the inverse operator of the last operator used [37]. Eventually, some operators may be self-adjunct. Adjunct operator T^* of an operator T is one that satisfies the condition of Equation 3.1:

$$\langle T(u), v \rangle = \langle u, T^*(v) \rangle. \tag{3.1}$$

Note that the adjunct operator is the operator's conjugated transposed matrix denoted by $T^* = \overline{T}^T$ [51]. Self-adjunct operator is one in which $T = T^*$, *i.e.*, the operator's adjunct is the operator itself. An example of such an operator is operator *Hadamard*, which will be described later in this section. Due to the fact that a state measurement leads the set of measured q-bits to assume irrevocably collapsed states, it hence also ends the reversibility of state. Note that, as any operation on a vector is equivalent to operating on the linear combination of the base of that same vector, a quantum operation on a certain state operates simultaneously on all q-bits of the system [47]. Throughout the remaining of this chapter, operations on one or more q-bits will be defined. Recall that a q-bit may be represented as a column vector of two elements, as seen in in Equation 2.2.

3.2 Tensor Product

When a quantum operation is performed on a state of n q-bits, with $n \geq 2$, a tensor product would be required, resulting in a column vector of 2^n rows [42]. A quantum operator on such a set of q-bits is a $2^n \times 2^n$ matrix. It is constructed from basic 2×2 quantum operators [56]. The tensor product between two matrices of any size is given by the product of each coefficient of the first matrix by its counterpart coefficient in the second matrix [41, 42]. Let $A_{m \times n}$ and $B_{p \times q}$ be two matrices, defined by Equation 3.2:

$$A = \begin{bmatrix} a_{11} & a_{12} & \dots & a_{1n} \\ a_{21} & a_{22} & \dots & a_{2n} \\ \dots & \dots & \dots & \dots \\ a_{m1} & a_{m2} & \dots & a_{mn} \end{bmatrix} \quad \text{and} \quad B = \begin{bmatrix} b_{11} & b_{12} & \dots & b_{1q} \\ b_{21} & b_{22} & \dots & b_{2q} \\ \dots & \dots & \dots & \dots \\ b_{p1} & b_{p2} & \dots & b_{pq} \end{bmatrix}.$$

(3.2)

The tensor product $C = A \otimes B$ is a matrix of $m.p$ rows and $n.q$ columns, as defined in Equation 3.3:

$$C = \begin{bmatrix} a_{11}b_{11} & a_{11}b_{12} & \dots & a_{11}b_{1q} & \dots & a_{1n}b_{11} & a_{1n}b_{12} & \dots & a_{1n}b_{1q} \\ a_{11}b_{21} & a_{11}b_{22} & \dots & a_{11}b_{2q} & \dots & a_{1n}b_{21} & a_{1n}b_{22} & \dots & a_{1n}b_{2q} \\ \dots & \dots & \dots & \dots & \dots & \dots & \dots & \dots & \dots \\ a_{11}b_{p1} & a_{11}b_{p2} & \dots & a_{11}b_{pq} & \dots & a_{1n}b_{p1} & a_{1n}b_{p2} & \dots & a_{1n}b_{pq} \\ \dots & \dots & \dots & \dots & \dots & \dots & \dots & \dots & \dots \\ a_{m1}b_{11} & a_{m1}b_{12} & \dots & a_{m1}b_{1q} & \dots & a_{mn}b_{11} & a_{mn}b_{12} & \dots & a_{mn}b_{1q} \\ a_{m1}b_{21} & a_{m1}b_{22} & \dots & a_{m1}b_{2q} & \dots & a_{mn}b_{21} & a_{mn}b_{22} & \dots & a_{mn}b_{2q} \\ \dots & \dots & \dots & \dots & \dots & \dots & \dots & \dots & \dots \\ a_{m1}b_{p1} & a_{m1}b_{p2} & \dots & a_{m1}b_{pq} & \dots & a_{mn}b_{p1} & a_{mn}b_{p2} & \dots & a_{mn}b_{pq} \end{bmatrix}.$$

(3.3)

The tensor product of n q-bits starts with the first 2 q-bits, then performs the tensor product between the result and the third q-bit until all tensor products including all the n q-bits have been performed. The result is a column vector with 2^n rows. To illustrate this computation, let $|\psi_0\rangle, |\psi_1\rangle, \dots, |\psi_n\rangle$ be n q-bits defined as in Equation 3.4:

$$|\psi_0\rangle = \begin{bmatrix} \alpha_0 \\ \beta_0 \end{bmatrix}, \; |\psi_1\rangle = \begin{bmatrix} \alpha_1 \\ \beta_1 \end{bmatrix}, \; |\psi_2\rangle = \begin{bmatrix} \alpha_2 \\ \beta_2 \end{bmatrix}, \; \dots, \; |\psi_n\rangle = \begin{bmatrix} \alpha_{n-1} \\ \beta_{n-1} \end{bmatrix}.$$

(3.4)

The tensor product of the first two q-bits produces a column vector of 4 rows, and is shown in Equation 3.5:

$$|\psi_{01}\rangle := |\psi_0\rangle \otimes |\psi_1\rangle = \begin{bmatrix} \alpha_0 \\ \beta_0 \end{bmatrix} \otimes \begin{bmatrix} \alpha_1 \\ \beta_1 \end{bmatrix} = \begin{bmatrix} \alpha_0 \alpha_1 \\ \alpha_0 \beta_1 \\ \beta_0 \alpha_1 \\ \beta_0 \beta_1 \end{bmatrix}. \tag{3.5}$$

Including the third q-bit, the tensor product of the three first q-bits produces a column vector of 8 rows, as shown in Equation 3.6:

$$|\psi_{012}\rangle = |\psi_{01}\rangle \otimes |\psi_2\rangle = \begin{bmatrix} \alpha_0 \alpha_1 \\ \alpha_0 \beta_1 \\ \beta_0 \alpha_1 \\ \beta_0 \beta_1 \end{bmatrix} \otimes \begin{bmatrix} \alpha_2 \\ \beta_2 \end{bmatrix} = \begin{bmatrix} \alpha_0 \alpha_1 \alpha_2 \\ \alpha_0 \alpha_1 \beta_2 \\ \alpha_0 \beta_1 \alpha_2 \\ \alpha_0 \beta_1 \beta_2 \\ \beta_0 \alpha_1 \alpha_2 \\ \beta_0 \alpha_1 \beta_2 \\ \beta_0 \beta_1 \alpha_2 \\ \beta_0 \beta_1 \beta_2 \end{bmatrix}^T. \tag{3.6}$$

This is done iteratively, considering a qubit at each time, until the last constructed column vector is multiplied with the nth q-bit, producing a column vector of 2^n rows, and the final result of the tensor product $|\psi_{012...n-1}\rangle = |\psi_{012...n-2}\rangle \otimes |\psi_{n-1}\rangle$ is obtained, as shown in

Equations 3.7:

$$|\psi_{012...n-1}\rangle = \begin{bmatrix} \alpha_0\alpha_1\alpha_2 \ldots \alpha_{n-3}\alpha_{n-2} \\ \alpha_0\alpha_1\alpha_2 \ldots \alpha_{n-3}\beta_{n-2} \\ \alpha_0\alpha_1\alpha_2 \ldots \beta_{n-3}\alpha_{n-2} \\ \alpha_0\alpha_1\alpha_2 \ldots \beta_{n-3}\beta_{n-2} \\ \ldots \\ \beta_0\beta_1\beta_2 \ldots \beta_{n-3}\beta_{n-2} \end{bmatrix} \otimes \begin{bmatrix} \alpha_{n-1} \\ \beta_{n-1} \end{bmatrix} = \begin{bmatrix} \alpha_0\alpha_1\alpha_2 \ldots \alpha_{n-2}\alpha_{n-1} \\ \alpha_0\alpha_1\alpha_2 \ldots \alpha_{n-2}\beta_{n-1} \\ \alpha_0\alpha_1\alpha_2 \ldots \beta_{n-2}\alpha_{n-1} \\ \alpha_0\alpha_1\alpha_2 \ldots \beta_{n-2}\beta_{n-1} \\ \ldots \\ \beta_0\beta_1\beta_2 \ldots \beta_{n-2}\beta_{n-1} \end{bmatrix}.$$

$$(3.7)$$

Tensor products are exploited in a similar manner to yield quantum operators to be used with quantum states of two or more quantum bits. The computation starts with the basic operator for one q-bit, which is a 2×2 matrix. For instance, let A and B be the matrix of two basic operators. The new operator C, which is used to operate on a state of 2 q-bits would be obtained as in Equation 3.8:

$$C = AB_\otimes = A \otimes B = \begin{bmatrix} a_1 & a_2 \\ a_3 & a_4 \end{bmatrix} \otimes \begin{bmatrix} b_1 & b_2 \\ b_3 & b_4 \end{bmatrix} = \begin{bmatrix} a_1.b_1 & a_1.b_2 & a_2.b_1 & a_2.b_2 \\ a_1.b_3 & a_1.b_4 & a_2.b_3 & a_2.b_4 \\ a_3.b_1 & a_3.b_2 & a_4.b_1 & a_4.b_2 \\ a_3.b_3 & a_3.b_4 & a_4.b_3 & a_4.b_4 \end{bmatrix}.$$

$$(3.8)$$

Therefore, an operator for n q-bits ($n \geq 1$) is constructed using the tensor product of the operator for $n-1$ q-bits with the basic quantum operator. So, this construction necessarily requires the tensor products between two basic quantum operators followed by the tensor products of an operator for m q-bits, $m = \{2, 3, \ldots, n-1\}$, with a basic quantum operator.

When it comes to a quantum operation on two or more q-bits, the tensor product between basic operators and that between the participating q-bits must be performed beforehand, and culminated by performing a matrix product between the yielded operator and the column vector thus obtained. It is noteworthy to point out that the tensor product has the property shown in Equation 3.9:

$$(A \otimes B).(u \otimes v) = (Au) \otimes (Bv), \tag{3.9}$$

wherein A and B are the quantum operators and u and v column vectors represent the quantum states used. This property is very important in order to build efficient simulators and emulators for quantum computing. For operations that neither cause entanglement nor deal with entangled q-bits, and as long as quantum state measurement is not required, using this property avoids building the operator for n q-bits, construction of the column vector representing the quantum state of n q-bits. We need only to enable the book keeping of the new states of the individual q-bits as modified by the basic operator, as long as they remain non-tangled. However, whenever a request of the measurement of the quantum state is issued, the tensor product among all q-bits would be executed.

3.3 Unitary Operators

There are seven main unitary quantum operators: *(i)* operator X, also termed operator NOT, rotates 180° around the x-axis, inverting the amplitudes associated with the base vectors. If applied to a q-bit in a collapsed state, it results in the opposite collapsed state, as in the case of a classic NOT gate [41]; *(ii)* operator Y rotates 180° around the y-axis, [41]; *(iii)* operator Z rotates 180° around the z-axis, [41]; *(iv)* operator I is the identity operator. It preserves the state of the q-bit it is applied to; *(v)* operator H, also termed operator Hadamard, transforms a q-bit in the collapsed state ($|0\rangle$ or $|1\rangle$) into an overlap of both states with equal amplitude [41, 42]; Operator H is a self-adjunct operator [56]; *(vi)* operator S implements the same transformation as operator H, but preserves the probabilities of obtaining either collapsed states, which does not happen with operator H [41]; *(vii)* operator T, also termed $\frac{\pi}{8}$ gate, performs a phase shift regarding ket "1" [41]. The definition of these operators and their application is shown in Table 3.1.

Table 3.1: Unitary quantum operators: definition and application.

Op	Symbol	Matrix	Linear definition	$\mathbf{O}\lvert 0\rangle$	$\mathbf{O}\lvert 1\rangle$
X	$-\boxed{\mathbf{X}}-$	$\begin{bmatrix} 0 & 1 \\ 1 & 0 \end{bmatrix}$	$\mathbf{X}\lvert v\rangle = \beta\lvert 0\rangle + \alpha\lvert 1\rangle$	$\lvert 1\rangle$	$\lvert 0\rangle$
Y	$-\boxed{\mathbf{Y}}-$	$\begin{bmatrix} 0 & -i \\ i & 0 \end{bmatrix}$	$\mathbf{Y}\lvert v\rangle = i(-\beta\lvert 0\rangle + \alpha\lvert 1\rangle)$	$-i\lvert 1\rangle$	$i\lvert 0\rangle$
Z	$-\boxed{\mathbf{Z}}-$	$\begin{bmatrix} 1 & 0 \\ 0 & -1 \end{bmatrix}$	$\mathbf{Z}\lvert v\rangle = \alpha\lvert 0\rangle - \beta\lvert 1\rangle$	$\lvert 0\rangle$	$-\lvert 1\rangle$
I	$-\boxed{\mathbf{I}}-$	$\begin{bmatrix} 1 & 0 \\ 0 & 1 \end{bmatrix}$	$\mathbf{I}\lvert v\rangle = \lvert v\rangle$	$\lvert 0\rangle$	$\lvert 1\rangle$
H	$-\boxed{\mathbf{H}}-$	$\begin{bmatrix} \frac{1}{\sqrt{2}} & \frac{1}{\sqrt{2}} \\ \frac{1}{\sqrt{2}} & -\frac{1}{\sqrt{2}} \end{bmatrix}$	$\mathbf{H}\lvert v\rangle = \frac{\alpha+\beta}{\sqrt{2}}\lvert 0\rangle + \frac{\alpha-\beta}{\sqrt{2}}\lvert 1\rangle$	$\frac{\lvert 0\rangle + \lvert 1\rangle}{\sqrt{2}}$	$\frac{\lvert 0\rangle - \lvert 1\rangle}{\sqrt{2}}$
S	$-\boxed{\mathbf{S}}-$	$\begin{bmatrix} 1 & 0 \\ 0 & i \end{bmatrix}$	$\mathbf{S}\lvert v\rangle = \alpha\lvert 0\rangle + i\beta\lvert 1\rangle$	$\lvert 0\rangle$	$i\lvert 1\rangle$
T	$-\boxed{\mathbf{T}}-$	$\begin{bmatrix} 1 & 0 \\ 0 & \varepsilon^{\frac{i\pi}{4}} \end{bmatrix}$	$\mathbf{T}\lvert v\rangle = \alpha\lvert 0\rangle + \frac{1}{\sqrt{2}}\beta(1+i)\lvert 1\rangle$	$\lvert 0\rangle$	$\frac{1+i}{\sqrt{2}}\lvert 1\rangle$

3.4 Binary Operators

There are two main binary quantum operators: *(i)* Controlled NOT, or simply CNOT, is one of the main operators in quantum computing, as it has the ability to entangle q-bits and its *modus operandi* is extensible to other quantum operators [41, 40]. This operator has two arguments: the control q-bit and the target q-bit. The CNOT operator inverts the state of the target q-bit when the control q-bit is in state 1. The magnitude of the inversion depends on the state of the control q-bit. It can be said that the CNOT operator is a NOT operator dependent on a second q-bit [39]. It can be constructed out of two chained H operators. The truth table for this port is shown in Figure 3.1(a), its matrix representation in Figure 3.1(b), its quantum symbol in Figure 3.1(c) and its operation in Figure 3.1(d); *(b)* Swap operator, as its name indicates swaps the contents of two q-bits [40]. It can be constructed out of three chained CNOTs or out of six chained H operators. The truth table for this port is shown in Figure 3.2(a), its matrix representation in Figure 3.2(b), the two commonly used quantum symbols in Figure 3.2(c) and its operation in Figure 3.3(d).

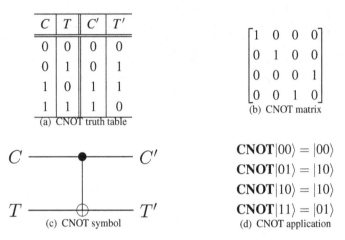

C	T	C'	T'
0	0	0	0
0	1	0	1
1	0	1	1
1	1	1	0

(a) CNOT truth table

$$\begin{bmatrix} 1 & 0 & 0 & 0 \\ 0 & 1 & 0 & 0 \\ 0 & 0 & 0 & 1 \\ 0 & 0 & 1 & 0 \end{bmatrix}$$

(b) CNOT matrix

(c) CNOT symbol

$\mathbf{CNOT}|00\rangle = |00\rangle$
$\mathbf{CNOT}|01\rangle = |10\rangle$
$\mathbf{CNOT}|10\rangle = |10\rangle$
$\mathbf{CNOT}|11\rangle = |01\rangle$

(d) CNOT application

Figure 3.1: Quantum operator CNOT: truth table, matrix, symbol and application to possible states.

3.5 Ternary Operators

There are two main ternary operators: *(i)* operator Fredkin or Controlled Swap, or simply CSWAP, changes the quantum state between two q-bits depending on a third q-bit, called the control q-bit. The truth table for this port is shown in Figure 3.3(a), wherein *A* and *B* are the q-bits that will have their states exchanged and *C* the control q-bit. A', B' and C' are the output states of these q-bits, respectively [23, 42]. Its matrix representation in Figure 3.3(b), its quantum symbol in Figure 3.3(c) and its application to possible quantum states in Figure 3.3(d); *(ii)* operator Toffoli uses three q-bits: two q-bits for control and a target q-bit. It reverses the state of the target q-bit whenever the two control q-bits are in state $|1\rangle$. The Toffoli operator can be

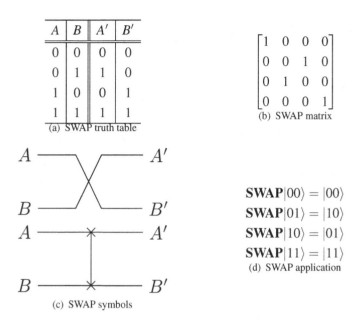

A	B	A′	B′
0	0	0	0
0	1	1	0
1	0	0	1
1	1	1	1

(a) SWAP truth table

$$\begin{bmatrix} 1 & 0 & 0 & 0 \\ 0 & 0 & 1 & 0 \\ 0 & 1 & 0 & 0 \\ 0 & 0 & 0 & 1 \end{bmatrix}$$

(b) SWAP matrix

(c) SWAP symbols

$$\textbf{SWAP}|00\rangle = |00\rangle$$
$$\textbf{SWAP}|01\rangle = |10\rangle$$
$$\textbf{SWAP}|10\rangle = |01\rangle$$
$$\textbf{SWAP}|11\rangle = |11\rangle$$

(d) SWAP application

Figure 3.2: Quantum operator SWAP: truth table, matrix, two commonly used symbols and applications to possible states.

seen as an extension of the CNOT port. It is considered universal and it is possible to build based on any other quantum operator [46]. The truth table for this operator can be found in Figure 3.4(a), wherein A and B are the control q-bits and C represents the target q-bit while A', B' and C' are the respective outputs. Figure 3.4(b) shows the operator matrix, Figure 3.4(c) the quantum used symbol for the Toffoli operator and Figure 3.4(d) its operation on possible quantum states.

3.6 Chapter Considerations

In this chapter, we explain the tensor product operation, which is the most important and most used mathematical operation during the execution of a quantum instruction. We show how this operation is used when more quantum bits become entangled with each other. We also explain that although the tensor product is not an end calculation in a quantum operation, it has significant importance due to the overall

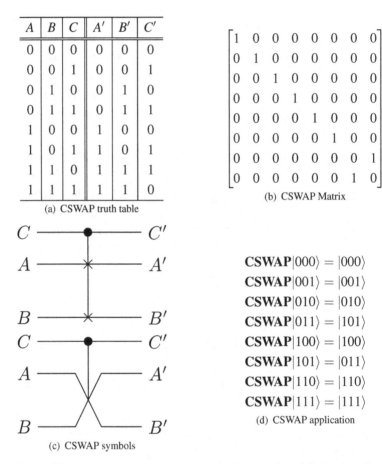

A	B	C	A'	B'	C'
0	0	0	0	0	0
0	0	1	0	0	1
0	1	0	0	1	0
0	1	1	0	1	1
1	0	0	1	0	0
1	0	1	1	0	1
1	1	0	1	1	1
1	1	1	1	1	0

(a) CSWAP truth table

$$\begin{bmatrix} 1 & 0 & 0 & 0 & 0 & 0 & 0 & 0 \\ 0 & 1 & 0 & 0 & 0 & 0 & 0 & 0 \\ 0 & 0 & 1 & 0 & 0 & 0 & 0 & 0 \\ 0 & 0 & 0 & 1 & 0 & 0 & 0 & 0 \\ 0 & 0 & 0 & 0 & 1 & 0 & 0 & 0 \\ 0 & 0 & 0 & 0 & 0 & 1 & 0 & 0 \\ 0 & 0 & 0 & 0 & 0 & 0 & 0 & 1 \\ 0 & 0 & 0 & 0 & 0 & 0 & 1 & 0 \end{bmatrix}$$

(b) CSWAP Matrix

$\mathbf{CSWAP}|000\rangle = |000\rangle$
$\mathbf{CSWAP}|001\rangle = |001\rangle$
$\mathbf{CSWAP}|010\rangle = |010\rangle$
$\mathbf{CSWAP}|011\rangle = |101\rangle$
$\mathbf{CSWAP}|100\rangle = |100\rangle$
$\mathbf{CSWAP}|101\rangle = |011\rangle$
$\mathbf{CSWAP}|110\rangle = |110\rangle$
$\mathbf{CSWAP}|111\rangle = |111\rangle$

(d) CSWAP application

(c) CSWAP symbols

Figure 3.3: Quantum operator CSWAP: truth table, matrix, symbol and application to possible states.

number of multiplications with complex numbers that it requires. We prove that this number increases exponentially with the number of quantum bits involved in the quantum state. We show that if not handled efficiently, it would require a lot of memory space, which makes the design emulate a quantum processor quite expensive to produce. We also define several common quantum operators with different arities: unitary, binary and ternary.

A	B	C	A'	B'	C'
0	0	0	0	0	0
0	0	1	0	0	1
0	1	0	0	1	0
0	1	1	0	1	1
1	0	0	1	0	0
1	0	1	1	0	1
1	1	0	1	1	1
1	1	1	1	1	0

(a) TOFFOLI truth table

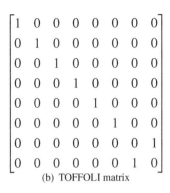

(b) TOFFOLI matrix

$$\mathbf{TOFFOLI}|000\rangle = |000\rangle$$
$$\mathbf{TOFFOLI}|001\rangle = |001\rangle$$
$$\mathbf{TOFFOLI}|010\rangle = |010\rangle$$
$$\mathbf{TOFFOLI}|011\rangle = |011\rangle$$
$$\mathbf{TOFFOLI}|100\rangle = |100\rangle$$
$$\mathbf{TOFFOLI}|101\rangle = |101\rangle$$
$$\mathbf{TOFFOLI}|110\rangle = |111\rangle$$
$$\mathbf{TOFFOLI}|111\rangle = |110\rangle$$

(d) TOFFOLI application

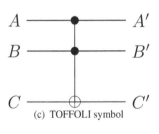

(c) TOFFOLI symbol

Figure 3.4: Quantum operator Toffoli: truth table, matrix, symbol and application to possible states.

In the next chapter, we present the macro-architecture of the proposed customizable quantum co-processor and how it communicates with the host processor in executing quantum instructions. We define the functionality of each of the main component of the quantum co-processor.

APPLICATIONS II

Chapter 4

Quantum Processor Macro-architecture

In this chapter, we present the overall hardware architecture of a quantum processor. First of all, we show its macro-architecture, explaining the role of each of its structuring components. Emulation of quantum processing requires primarily an efficient way to store and retrieve quantum information. The quantum processor employs different purpose memories. There is the quantum state memory, which stores information about the possible states of the quantum bits; the quantum operator memory, which stores the quantum operators in their basic form; the scratch memory, which allows storing intermediate results regarding the computation of higher-dimensionality operators form the basic ones, when these are required. For each of these memories, we describe the format used to organize the required information and show simulation results about its usage both in read and write cycles.

4.1 Introduction

The proposed co-processor implements a quantum machine capable of performing certain quantum operations on a set of q-bits that form the machine quantum state. It is an isolated system that interacts with a host processor through communication channels. The main processor sends properly formatted commands that invoke quantum operations on q-bits, to manipulate the quantum state. Such a quantum state remains hermetic in the quantum machine until the co-processor is commanded to read its state, resulting in the assumption of a collapsed state for each q-bit. The co-processor is designed to be able to perform quantum operations on 1 or more q-bits of the machine state based on either basic quantum operators or operators built at runtime, to meet the main processor's requests. Once the execution of the quantum instruction is complete, it rests waiting for a new instruction. It may return the up-to-date machine quantum state when requested. The basic operators implemented are those described in Chapter 3, such as H, X, Y, Z, T, S, I and CNOT. Any other quantum operator can also be executed, given that the respective basic operator matrix is uploaded into the memory of the co-processor.

Figure 4.1 illustrates the communication of the main processor (PROC) and quantum co-processor (COPROC) through a half-duplex channel. The channel configuration is used because it considers the sequential characteristic of a quantum algorithm due to the potential dependence of the result of the previous operation on one or more q-bits. The full-duplex configuration would also be possible, given the required control related to data dependency. Quantum operations are requested by PROC using descriptive blocks that specify the basic quantum operation and the target q-bit. If the operation is over two or more q-bits, two or more descriptors are required to specify the operation and the target q-bits. Of course, when requested by PROC, COPROC reads the quantum state. This will be detailed in Chapter 5.

The macro-architecture of the quantum processor is shown in Figure 4.2. Unit UCO represents the control unit of COPROC. It manages, through a micro-program and some specific components, the operation of the data-path of the architecture. The control unit records, decodes and interprets the quantum descriptors about the quantum operation code and the target q-bit (s). The co-processor

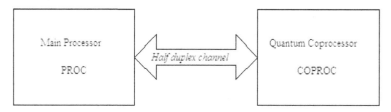

Figure 4.1: Communication between the main processor and the quantum co-processor.

disposes of three memories: *(i)* memory MQS stores the machine quantum state; *(ii)* memory MOP is used to keep the coefficients of the basic operators; *(iii)* memory MSC is the scratch memory and stores the coefficients of the quantum operators for two or more q-bits, which are calculated from the basic quantum operators. Unit UCA represents the calculation unit of COPROC. It basically performs tensor and matrix products. The tensor product is required between q-bits, between basic operators and between the calculated operator and basic operator while the matrix product is required as an operator and a q-bit. Unit UMS represents the measurement unit. It performs the quantum state measurement with the help of the RNG component, which is a pseudo-random number generator. When requested by PROC, UCO provides the measurement result or the probabilities associated with the possible states.

The interaction between the units and memories that compose the co-processor is done via data buses, address buses and/or control buses. The five data buses are: unidirectional operator data bus (ODATA) of 128 bits, which provides the calculation unit with the coefficients of the matrices representing the quantum operators; bidirectional scratch data bus (SDATA) of 128 bits, which allows the exchange of intermediate results during the computation of the tensor products of operators of two or more bits; bidirectional quantum bus data (QDATA) of 128 bits, which allows the exchange of states of the q-bits addressed by the quantum instructions as sent by the main processor; bidirectional measurement data bus (MDATA) of 128 bits, which allows the exchange of the states of the observed quantum bits before and after measurements; unidirectional random number bus (RANDOM) of 32 bits, which forwards the generated

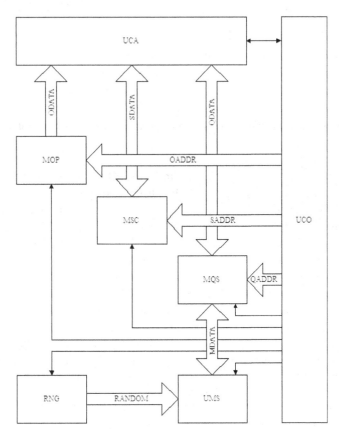

Figure 4.2: Co-processor macro-architecture.

pseudo random number necessary for q-bit measurement. Moreover, the co-processor includes three address buses, one for each required memory: OADDR whose number of bits is dependent on the number and composition of the stored quantum operator as explained later; QADDR and SADDR, whose number of bits coincides with the maximum number of q-bits of the machine quantum state. Finally, there are six control buses, one of each functional unit of the macro architecture. All the control units are unidirectional originating from the control unit, except that between UCO and UCA, which is bidirectional, as it will be detailed later when explaining the UCA operation.

Before, we get into further details, it is fundamental to state that the proposed co-processor design, described in the following sections, is customizable in terms of four main parameters. These are the maximum number of q-bits that the co-processor can have in its state, denoted by nq; the maximum number of q-bits that the co-processor can operate on simultaneously, denoted by mq; the total number of basic operators, denoted by nop; and for each available basic operator o, the number of q-bits it operates on, denoted by noq_o.

4.2 Quantum State Memory

The quantum state memory (MQS) is a read-write dual-port memory, with a restriction for simultaneous read and write operation at the same address. It consists of two parts: the MQB for the q-bit state memory and the MQC for the q-bit control memory. Both have the same number of addresses, and their contents are linked together for each address, as if one were an extension of the other. This separation is due to the possibility that only one of these memories could be updated in certain situations during quantum instruction executions. The number of addresses in MQS meets the need to represent all possible states of the quantum machine. Recall that nq denotes the maximum number of q-bits of the machine quantum state. So, the first nq addresses are reserved for storing the kets of the q-bits, not yet entangled or already entangled. The remaining $2^{nq-1} - nq$ addresses are destined to store the entries of the column-vector that represent a set of entangled q-bits, thus complementing the set of coefficients of the column-vector of the nq initial addresses, required for the entangled q-bits. It is noteworthy to point out that with this organization, the machine can handle at most nq q-bit entanglements. The memory MQS is organized in such a way as to hold at each address the two kets of each non-entangled q-bit of the machine quantum state, augmented by the required data to describe the amplitudes of each possible state relative to the set of two or more entangled q-bits. At each address a of this memory, the data is divided into: *(i)* the two coefficients (real and imaginary parts) for both kets of the q-bit are stored in MQB at address a, formatted as shown in Figure 4.3; *(ii)* the addresses of the first and next q-bit of the entangled q-bit list of which the q-bit at address a is part of, are stored in MQC at address a, formatted as shown in Figure 4.4.

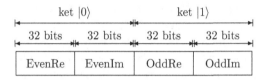

Figure 4.3: Word format of memory MQB.

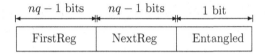

Figure 4.4: Word format of memory MQC.

Figure 4.5, shows the MQS logic block. The input and output data, which comes and goes out from bus QDATA, is organized into four parts: the real and imaginary of the complex numbers representing the amplitudes of kets $|0\rangle$ (even) and $|1\rangle$ (odd) of the q-bits. This data organization is useful to the design of unit UCA, as it will be justified later. It is noteworthy to point out that for all the logic blocks presented herein, the input pins regarding data, address and control signals are shown in black, white and gray respectively, while the output pins regarding data and control signals are shown as hatched and dotted respectively. There are no output address signals in all the proposed designs.

The information about complex numbers representing the amplitude of the kets is stored in memory MQB, whose logic block is shown in Figure 4.6. The non-entangled condition of a q-bit is potentially transient in a quantum algorithm. Thus, the address originally assigned to the non-entangled q-bit, after entanglement, will be used to store the first two coefficients of the vector-column representing the set of entangled q-bits to which it belongs, one relative to ket $|0\rangle$ (even) and the other to ket $|1\rangle$ (odd). These odd and even references will be useful in understanding the design of unit UCA, described later. During the usage of the quantum machine, each posi-

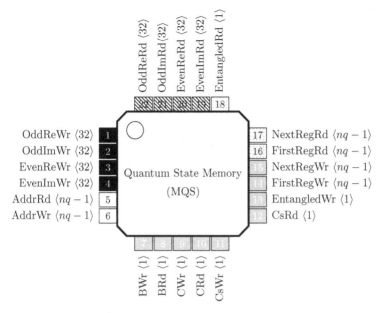

Figure 4.5: Logic block MQS for the quantum state memory.

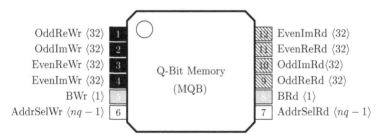

Figure 4.6: Logic block MQB for the q-bit memory.

tion in memory MQB is initialized with the tuple (1.0, 0.0, 0.0, 0.0), representing coefficients $1 + 0i$ and $0 + 0i$, according to the format of Figure 4.3. So, we have the ket $|0\rangle$ represented with the value 1.0 in the real part and 0.0 in the imaginary part, while ket $|1\rangle$ with value 0.0 in both imaginary and real parts [37]. Note that all data is kept in IEEE754 standard floating number format. The time diagram for the initialization operation is shown in Figure 4.7.

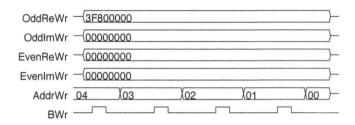

Figure 4.7: Time diagram of MQB initialization.

OddReRd				65 85	
OddImRd				65 85	
EvenReRd				65 85	
EvenImRd				65 85	
BRd					
AddrSelRd	0	1 2 3		0 1 2 3	
OddImWr	0		65 85		
OddImWr	0		65 85		
EvenReWr	0		65 85		
EvenImWr	0		65 85		
BWr					
AddrSelWr	0		1		

Figure 4.8: Time diagram of simultaneous read and write operations on a non-initialized MQB.

Figure 4.8 shows the time diagram of simultaneous reads and writes on a non-initialized memory MQB. Initially, addresses 0 to 3 are not initialized. The control signals are active when they are high. The read signal is kept active throughout the simulation interval to demonstrate the correctness of the memory operation when writes occur. Note that the write operation of value 65, as the write control signal is active, performed at address 1 and then value 85 is written at address 2. Observe that these values are obtained when the corresponding addresses are read. Address 3, having not been subject to a write operation at any time, remained unchanged.

To be able to simulate q-bit entanglement, we associate each q-bit of the quantum state with the references to the addresses of the first and next q-bit of the sequence of entangled q-bits it is part of, if

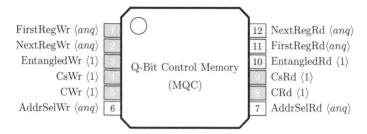

Figure 4.9: Logic block MQC for the q-bit control memory.

any, as shown in Figure 4.4. Note that whether a q-bit is entangled is indicated with a entanglement bit, which is part of the stored word, set to 1 in the positive case. In case of a non-entangled q-bit, these addresses are set to the address of the q-bit itself. This information is stored in memory MQC, whose logic block is shown in Figure 4.9. Note that for both memory MQB and MQC, parameter *anq* represents the number of bits required to select the addressed positions in memories MQB and MQC. Actually, the result of the binary decodification of the address of memory MQS is fed as read/write selector to both memories MQB and MQC. So two binary address decoders are used to decode $nq - 1$ bits into the corresponding one-hot code of the address. So, we have $anq = 2^{nq-1}$ bits.

The need to form a partially chained list, and not a completely chained one is because to be able to perform the required operations to simulate q-bit entanglement, one does not need to know the q-bit immediately preceding the currently considered one, since any quantum operation, including entanglement, that is done on a set of q-bits must always start with the first q-bit of the sequence of q-bits. Recall that a sequence of e entangled q-bits require the memory space of 2^{e-1} locations, *i.e.*, $128 \times 2^{e-1}$ bits, because they are the result of a tensor product between all e q-bits of the sequence and a single memory word, we store both the even and odd coefficients. As a quantum operation occurs over a set of q-bits, it is necessary to know the address of the first one, since the quantum operation, which is performed as a matrix product, will start with the first row of the column-vector representing the entangled q-bits. As each address contains information corresponding to two rows of the column vector of e q-bits, the position number to be read in a quantum operation will be 2^{e-1}, considering that each MQB address contains four coefficients. When performing a reset of the quantum machine, every address a in MQC will have tuple $(a, a, 0)$ as its initial setting.

The total number of bits in quantum state memory MQS can be evaluated as the sum of those of memories MQB and MQC. Both memories have the same addressable space, extending from 0 to 2^{nq-1}. However, the word size is different. In the case of MQB, each position has a fixed size of 128 bits while in the case of MQC, it has a variable, depending on the number of q-bits of the co-processor. Each MSC word has $2nq - 1$. Hence, for the total of nq q-bits, the size of memory MQS in terms of bits can be defined as in Equation 4.1:

$$Size_{MQS} = 2^{nq-1}(2nq - 127).$$ (4.1)

Figure 4.10 shows the time diagram of simultaneous reads and writes on a non-initialized memory MQC. Registers *CtrlRegRd* and *CtrlRegWr* represent the contents of the words in MCQ, and are composed of the three fields: $\langle FirstRegRd, NextRegRd, EntangledRd \rangle$ and $\langle FirstRegWr, NextRegWr, EntangledWr \rangle$, respectively. Addresses *AddrSelRd* and *AddrSelWr* determine the read and write address, respectively. Signals *CsRd* and *CsWr* are read and write chipSelect, respectively. Signals *CRd* and *CWr* enable reading and writing cycles, respectively. Initially, observe that the memory locations 0 to 4 are without data recorded in the *CtrlRegRd*. Then, the writing of

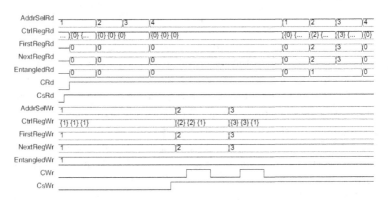

Figure 4.10: Time diagram of simultaneous read and write operations on a non-initialized MQC.

value 0x1 in the fields *FirstRegWr*, *NextRegWr* and *EntagledWr* of address 1, value 0x2 in the aforementioned fields of address 2 and value 0x3 in the same fields of address 3, except of course for field *EntangledWr*, which remains with value 0x1 for all cases. Also, observe the enabling of signals *CsWr* and *CWr*. Next, the reading of addresses 0 to 3, shows that the contents are saved only at addresses 2 and 3, when signals *CWr* and *CsWR* are enabled.

4.3   Quantum Operator Memory

The quantum operator memory MOP is a read-only memory that stores the coefficients of the basic quantum operators. A basic quantum operator is represented by either a 2×2 or 2×2 square matrix. Its words are organized as for memory MQB (see Figure 4.3). So, each address in MOP comprises the two coefficients of a given row of the operator matrix, termed odd and even columns. Therefore, two contiguous MOP addresses store the coefficients of a basic 2×2 quantum operator. So we have bits $0 \dots 31$, $32 \dots 63$, $64 \dots 95$ and $96 \dots 127$ corresponding to the coefficient real part of the odd column, coefficient imaginary part of the odd column, coefficient real part of the even column and coefficient part imagery of the even column. Figure 4.11 shows the logic block of this memory, wherein parameter *aop*, denotes the number of required bits to address memory MOP, which is defined as in Equation 4.2:

$$aop = \left\lceil \log_2 \left(\frac{1}{2} \sum_{o=1}^{nop} 4^{noq_o} \right) \right\rceil . \tag{4.2}$$

It coincides with half the total number of complex coefficients of the implemented quantum operators by the quantum co-processor. Recall that parameter *nop* represents the number of basic operators and parameter noq_o represents the number of q-bits operators o required to be operated.

This arrangement of storing of two complex numbers by MOP address is compatible with the of kets coefficients arrangement in memory MQB. It allows us to design an efficient tensor and matrix multipliers, as it will be explained in Chapter 5. In this implementation we include operators: I, X, Y, Z, H, S, T and CNOT. Note that for

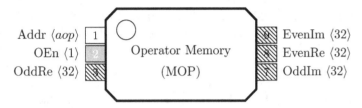

Figure 4.11: Logic block MOP for the quantum operator memory.

all other existing quantum operators, the corresponding matrix can be obtained from the implemented ones. Quantum operators are referenced in the quantum program by a unique code that is associated with the initial address of the range reserved for the respective quantum operator. Considering the list of implemented operators, instead of using 3 bits to code the 8 operators and an extra lookup table that allows us to have access to the starting address of the addressed operator matrix in MOP, we use only four bits that indicate the address of the word regarding the requested operator, as stored in memory MOP. Table 4.1 shows the binary codes of the implemented quantum operators, the number of q-bits the operators o require (noq_o) as well as the total number of coefficients of the operator o ($noc_o = 4^{noq_o}$).

Table 4.1: Codification of the implemented quantum operators.

Op	Code	Noq_o	Noc_o	Address
I	0000	1	4	0–1
X	0010	1	4	2–3
Y	0100	1	4	4–5
Z	0110	1	4	6–7
H	1000	1	4	8–9
S	1010	1	4	10–11
T	1100	1	4	12–13
CNOT	1110	2	16	14–21

The total number of bits in operator memory MOP can be evaluated as the sum of those of the space required to hold all permitted operators by the co-processor. Hence, for the total of *nop* coefficients all basic operators considered, the size of memory MOP in terms of bits can simply be defined as in Equation 4.3:

$$Size_{MOP} = 128 \times \exp\left(\left\lceil \log_2\left(\frac{1}{2}\sum_{o=1}^{nop} 4^{noq_o}\right)\right\rceil\right) \qquad (4.3)$$

Figure 4.12 shows the time diagram of a reading operation of the MOP at addresses 0 and 1, respectively containing the coefficients of the first and second entries of the identity operator. The hexadecimal values 0x3F800000 and 0x00000000 correspond to values 1.000 and 0.000 in the IEEE754 format. The real part of the coefficients at addresses 0 (*i.e.*, entries [0,0] of the operator matrix) and 1 (*i.e.*, entries [1,1] of the operator matrix) have a value of 1.000, while the imaginary part of the other two coefficients as well as the real part of the coefficient (0,1) and (1,0) are equal to 0.000.

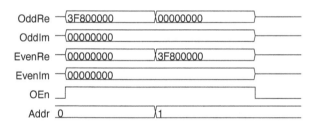

Figure 4.12: Time diagram for the read operation of MOP.

4.4 Scratch Memory

The scratch memory (MSC) is used to store the coefficients of quantum operators for 2 or more q-bits, built at runtime, by means of a tensor product of basic quantum operators. The MSC is a read-write dual-port memory, with restrictions for simultaneous reading and writing at the same address. The corresponding logic block is shown in Figure 4.13.

Memory MSC's word is structured as that of memory MQB, as shown in Figure 4.3. Analogous to memory MOP, each address in MSC comprises two coefficients of an operator matrix row: one for

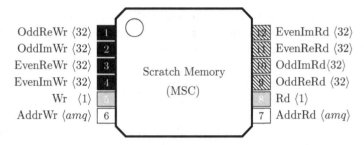

Figure 4.13: Logic block MSC for the scratch memory.

the odd column and the other for the even column, supplying one of the inputs of the 2 multipliers of complex numbers embedded into the calculation unit UCA, and used during the computation of both tensor and matrix products. The total number of addresses in MSC, denoted by a, is dependent on the maximum amount of q-bits that could be operated simultaneously in the quantum machine, denoted mq. So, we have $a = 4^{mq}/2 = 2^{2mq-1}$ and thus $amq = 2mq - 1$. Knowing that mq is at most equal to the total number of q-bits in the quantum machine, denoted by nq. Then at most we have $amq = 2nq - 1$. Memory MSC is only used during the construction and temporary storage of the quantum operator of more than two q-bits, according to the instructions extracted from the quantum program in execution.

The total number of bits in operator memory MSC can be evaluated as the sum of those of the space required to hold the largest operator the co-processor can handle. This is defined by parameter mq. Hence, the size of memory MSC in terms of bits and its upper bound can be defined as in Equation 4.4:

$$Size_{MSC} = 128 \times 2^{2mq-1} \qquad (4.4)$$

and its upper bound is as defined in Equation 4.5:

$$\overline{Size}_{MSC} = 128 \times 2^{2nq-1}, \qquad (4.5)$$

Figure 4.14 shows the time diagram of simultaneous read and write operations in MSC. Initially, the memory positions 0 to 3 are read and show 0 as corresponding contents. Then, the addresses 1 and 2 are written with arbitrary values 15 and 25 respectively. Finally, the addresses 0 to 3 are read again, but now show the modified contents of positions 1 and 2.

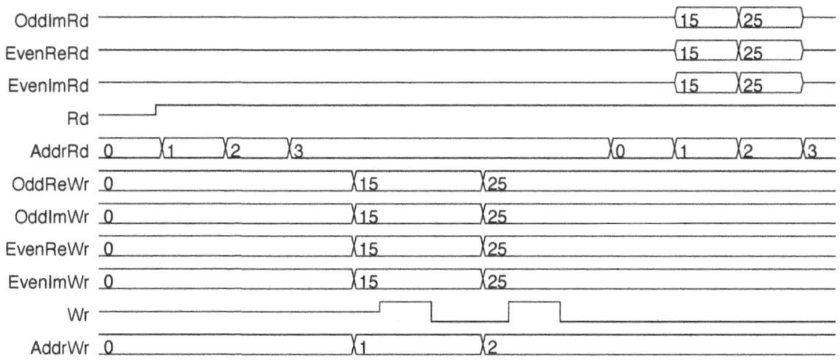

Figure 4.14: Time diagram for the read operation of MSC.

4.5 Chapter Considerations

In this chapter, the macro-architecture is presented and the functionality of the included components are explained. We show that the co-processor requires three kind of memories. The first one is used to store the quantum state (MQS). The second one is used to store the coefficients of the basic quantum operators (MOP). The third one is used to store the coefficients of the constructed quantum operator (MSC). The role and the word structure of each of these memories are described and motivated. Due to their complexity, the functional unit shown in the macro-architecture calculation unit (UCA), the control unit (UCO) and the measurement unit (UMS) together with the pseudo-random number generator (RNG) require dedicated chapters.

In the next chapter, we present the micro-architecture of the calculation unit and show how it computes matrix and tensor products efficiently, requiring only 4 complex number multipliers that operate in the pipeline.

Chapter 5

Calculation Unit Micro-architecture

In this, chapter, we first give an overall view regarding the function that must be executed by the calculation unit. We present and explain the underlying micro-architecture of this functional unit. The task of this unit is to allow an efficient computation of tensor and matrix products, which are required by any quantum information manipulation made by the processor. It also allows the computation of complex number summations. For this purpose, the unit implements efficient parallel complex number multiplications, which form the foundations of tensor and matrix products. After that, we explain and give illustrative examples on how the calculation unit proceeds and controls the computation of the tensor product of quantum bits and registers. Then, we do the same for the tensor products of operators, required to obtain an operator of high dimensionality from the basic one pre-stored in the quantum operator memory. Last but not the least, we explain how the calculation unit proceeds to provide a required matrix product.

5.1 Introduction

The Calculation Unit UCA of the macro-architecture is responsible for multiplication of complex numbers, necessary for the execution of the tensor product of quantum operators, tensor product of q-bits, matrix product of operators and quantum register of q-bits as well as the sum of complex numbers. Unit UCA can receive data for memories MSC, MQS and MOP. As for the other main component of the macro-architecture, its operation is managed by control unit UCO.

The micro-architecture of unit UCA is depicted in Figure 5.1. It is connected to two input and two output data buses of 128 bits each. The input data bus forwards the coefficient pair of the column vector representative of the q-bit or set of q-bits, from memory MQS, while the second input data bus pairs the coefficient pair from either memory MOP or MSC. Recall that MOP provides the coefficients of the basic quantum operators, while MSC provides the coefficients of an operator built for 2 or more q-bits. Unit UCA connects to output data buses for writing in MSC and MQS separately. The data intended for the MSC writing data bus is that about the tensor product being calculated (intermediate results) while that for the MQS writing data bus is about the final result of the tensor product of q-bits or of the matrix product between the quantum operator and a q-bit or a register of q-bits.

In the architecture of Figure 5.1, registers RMC1 and RMC2 store the coefficient of the q-bit or the operator in the course of multiplication. Registers RMCTP1 and RMCTP2 record the coefficients of the second row of basic quantum operators when computing a tensor product. Registers $RMCEX1_{1...n}$ and $RMCEX2_{1...n}$ represent a set of n registers designed to store the data located at the positions of MQS that are involved in the quantum operation before the operation takes place. The number of this kind of registers is defined by, $n = 2^p - 2$, where p is the maximum amount of q-bits on which an operator can act. Registers RMCTP11 and RMCTP21 store coefficients of the first row of the basic quantum operator while RMCTP12 and RMCTP22 store those of the second row.

Components MULTC1 and MULTC2 are multipliers of complex numbers while components SUMC1 and SUMC2 are adders of complex numbers. The first adder is used to sum up 2 partial products in a

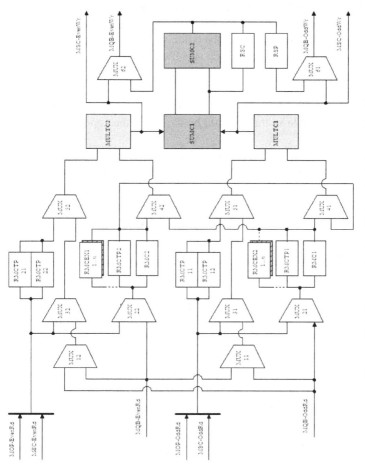

Figure 5.1: Micro-architecture of the calculation unit UCA.

matrix multiplication between an operator and a set of q-bits, and the second is responsible for the accumulation of the partial products of an operator matrix row with the column vector represented by 1 q-bit or more entangled q-bits. This accumulation is performed by repetitively using complex number register RSC. Register RSP holds the intermediate complex number regarding the sum of computed partial products to be stored in MSC, until the corresponding final result becomes available so that it could be written at the appropriate address of MQS.

The complex number multipliers (MULTC1 and MULTC2) are based on simple precision floating-point arithmetic unit IPs (FPU) [26]. The unit is customizable so it can be tailored according to the required operation. We use 4 multipliers, 1 adder and 1 subtractor. Let us consider 2 complex numbers as ordered pairs (A, B) and (C, D), where A and C denote the coefficients of the real part while B and D denote the coefficients of the imaginary part. The multiplication of these two complex numbers would result into a complex number represented by the pair $(AC - BD, AD + BC)$, with $AC - BD$ its real part and $AD + BC$ its imaginary part. The 4 multiplications to obtain the partial products AC, AD, BC and BD are done first and in parallel. When the products are ready, the sum $AD + BC$ and the difference $AC - BD$ may occur concurrently.

The micro-architecture of the complex number multiplier is shown in Figure 5.2. As the used FPU operates continuously, not depending on a specific trigger, the complex number multiplier can be supplied in a serial manner with data and the results will be made available sequentially. Each used FPU considers the data present in its input pins at the clock rising transition and the result is also sampled at a clock rising transition. The number of clock cycles required to get the correct result is determined by the FPU latency parameter. For multipliers, this parameter is set to 3 clock cycles while for adder and substractors, it is set to 2. It is noteworthy to point out that the used clock signals are different, so that the products could be used by the adder/subtractor before the next transition of the multiplier clock signal. If such an approach were not used, a delay of 1 clock cycle would be introduced during every product of complex numbers, which would result in a considerable delay when the quantum operation involves several q-bits. The total amount of complex

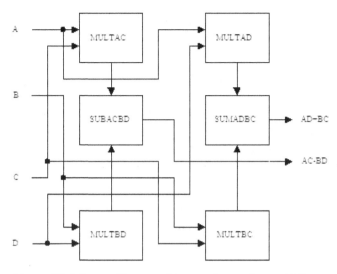

Figure 5.2: Micro-architecture of the complex number multiplier.

number multiplications in a quantum operation that involves $n \geq 2$ q-bits sums up to $2^n + 4^n + 4^n = 2^n + 4^{n+1} = 2^n + 2^{2(n+1)} = 2^{3n+2}$, which encompass the tensor product of the n q-bits augmented by the tensor product to yield the required quantum operator from basic ones and further augmented with the matrix multiplication between the thus obtained quantum operator by the result of the tensor product of the q-bits involved in the operation. For instance, for a single quantum operation on 3 q-bits, the co-processor would spend 6,144 clock cycles doing multiplications of floating-point numbers.

5.2 Tensor Product of Q-bits

The construction of a quantum register starts by the tensor product of the q-bits specified in the first two instructions of the quantum operation. The q-bit that is specified in the third instruction would be tensorally multiplied with the register of two q-bits that was prepared in the previous iteration. For tensor products between four or more q-bits, we proceed analogously, multiplying the obtained column vector by the q-bit defined in the next instruction. The number of rows of the matrix representing the n q-bit tensor product is 2^n, requiring $2^n - 1$ positions of the memory MQS. Recall that the tensor product between q-bits extends the entanglement condition to all participating q-bits. Memory MQB, initially comprising the two coefficients of the 2×1 column vector that it represents, would after this operation occupy $2^n - 1$ addresses (*i.e.*, 2^n coefficients), n denoting the number of entangled q-bits involved in the operation. Memory MQC stores the pointers that trace the order of the data of each participating MQB address. For instance, if q-bits 0, 1 and 3 participate in a quantum operation, with q-bits 0 and 3 being entangled, at the end of the operation, address 1 would also store the entanglement mark ($Em = 1$). MQB Addresses 0, 1 and 3 would be necessary but not sufficient to accommodate the 8×1 column-vector result of the tensor product, because an extra address (the fourth address in this group) would be required to store the last two coefficients of the resulting column vector. Furthermore, if q-bit 2 gets entangled with q-bits 0, 1 and 3, the resulting 16×1 column-vector would require 3 extra entries in MQB and MQC. In general, the total number of required extra entries in MQS is $2^{n-1} - n$, wherein n represents the number of entangled q-bits. The contents dynamics of memories MQB and MQC regarding the aforementioned example are illustrated in Figure 5.3.

5.3 Tensor Product of Operators

Recall that a quantum operator that operates on $q \leq 1$ q-bits simultaneously, has 2^q rows and columns. When executing the operator tensor product for 2 q-bits, a row of the basic operator coefficients (2×2) is read each time from memory MOP. The addresses of memory MOP that must be read are provided by unit UCO according to the quantum instruction sent by the main processor. The tensor product of two operators starts by loading the coefficients of the first row of the first operator in registers RMC1 and RMC2, followed by loading the coefficients of the second line in registers RMCTP1 and RMCTP2, via multiplexers MUX21 and MUX22. Next and analogously, the coefficients of the first row of the second operator are loaded in registers RMCTP11 and RMCPT21 and those of the second row in registers RMCPT12 and RMCPT22. Then, they are sent to multipliers MULTC1 and MULTC2 inputs, enabling outputs from

Address	MQB	MQC			MQB	MQC		
0	$(1,0,0,0)$	0	0	0	$(\alpha_0,\beta_0,\gamma_0,\delta_0)$	0	3	1
1	$(1,0,0,0)$	1	1	0	$(1,0,0,0)$	1	1	0
2	$(1,0,0,0)$	2	2	0	$(1,0,0,0)$	2	2	0
3	$(1,0,0,0)$	3	3	0	$(\alpha_3,\beta_3,\gamma_3,\delta_3)$	0	3	1
4	$(-,-,-,-)$	4	4	0	$(-,-,-,-)$	4	4	0
5	$(-,-,-,-)$	5	5	0	$(-,-,-,-)$	5	5	0
6	$(-,-,-,-)$	6	6	0	$(-,-,-,-)$	6	6	0
7	$(-,-,-,-)$	7	7	0	$(-,-,-,-)$	7	7	0

(a) Initial state (b) Q-bit 0 with 3

Address	MQB	MQC			MQB	MQC		
0	$(\alpha_0',\beta_0',\gamma_0',\delta_0')$	0	3	1	$(\alpha_0'',\beta_0'',\gamma_0'',\delta_0'')$	0	3	1
1	$(1,0,0,0)$	1	1	0	$(\alpha_1,\beta_1,\gamma_1,\delta_1'')$	0	5	1
2	$(\alpha_2,\beta_2,\gamma_2,\delta_2)$	0	4	1	$(\alpha_2',\beta_2',\gamma_2',\delta_2')$	0	4	1
3	$(\alpha_3',\beta_3',\gamma_3',\delta_3')$	0	2	1	$(\alpha_3'',\beta_3'',\gamma_3'',\delta_3'')$	0	2	1
4	$(\alpha_4,\beta_4,\gamma_4,\delta_4)$	0	4	1	$(\alpha_4',\beta_4',\gamma_4',\delta_4')$	0	1	1
5	$(-,-,-,-)$	5	5	0	$(\alpha_5,\beta_5,\gamma_5,\delta_5)$	0	6	1
6	$(-,-,-,-)$	6	6	0	$(\alpha_6,\beta_6,\gamma_6,\delta_6)$	0	7	1
7	$(-,-,-,-)$	7	7	0	$(\alpha_7,\beta_7,\gamma_7,\delta_7)$	0	7	1

(c) Q-bit 2 with $|\psi_{03}\rangle$ (b) Q-bit 1 with $|\psi_{032}\rangle$

Figure 5.3: Illustration of the dynamics of memories MQB and MQC during q-bits entanglements.

the involved registers, generating the operator coefficients for 2 q-bits given 2 basic 2×2 operators $A = [a_{ij}]$ and $B = [b_{ij}]$ for $1 \le i, j \le 2$. Table 5.1 shows the names of the registers that provide for the associated product computations for each of the 4 rows of the resulting quantum operator.

Table 5.1: The used registers to provide the operand for the computation of the tensor product of 2 q-bits.

#Row	Register 1	Register 2	Register 3	Product 1	Product 2
1	RMC1	RMCPT11	RMCPT21	$a_{11}.b_{11}$	$a_{11}.b_{12}$
	RMC2	RMCPT11	RMCPT21	$a_{12}.b_{11}$	$a_{11}.b_{12}$
2	RMC1	RMCPT12	RMCPT22	$a_{11}.b_{21}$	$a_{11}.b_{22}$
	RMC2	RMCPT12	RMCPT22	$a_{12}.b_{21}$	$a_{21}.b_{22}$
3	RMCPT1	RMCPT11	RMCPT21	$a_{21}.b_{11}$	$a_{21}.b_{12}$
	RMCPT2	RMCPT11	RMCPT21	$a_{22}.b_{11}$	$a_{22}.b_{12}$
4	RMCPT1	RMCPT12	RMCPT22	$a_{21}.b_{21}$	$a_{21}.b_{22}$
	RMCPT2	RMCPT12	RMCPT22	$a_{22}.b_{21}$	$a_{22}.b_{22}$

In order to compute the tensor product of 3 q-bits, we start with the tensor product of the two basic operators, whose tensor product of the coefficients of the third operator must be performed. The operator built for 2 q-bits has 16 coefficients, consuming 8 positions in memory MSC. Recall that the operator regarding n q-bits has 4 times more coefficients than the operator for $n - 1$ q-bits, *i.e.*, has 2 times more addresses occupied in MSC. This memory is dual-port to allow the tensor product of operators to occur continuously. However the same memory position cannot be read and/or written simultaneously. For this purpose, the construction of the tensor product for 3 or more q-bits starts with the last coefficient of the original operator. In this case, the operator for 2 q-bits is traversed in the reverse sense: going through each coefficient from the last to the first column and from the last to the first row. Such a reverse order for reading the coefficients is adopted to prevent conflict in the addressing memory MSC, allowing for a more efficient computation of the final result.

During the computation of the tensor product operator for 3 q-bits, the co-processor must have 16 registers: 8 for the odd column coefficients (RMC1, RMCTP1, RMCEX1$_{1...6}$) and 8 for the operator pair (RMC2, RMCPT2, RMCEX2$_1$ ... 6) for 2 q-bits, to be registered in the following order:

■ $a_{22}b_{21}$ and $a_{22}b_{22}$ are loaded in RMC1 and RMC2;

■ $a_{21}b_{21}$ and $a_{21}b_{22}$ in RMCPT1 and RMCPT2;

- $a_{22}b_{11}$ and $a_{22}b_{12}$ RMCEX1$_1$ and RMCEX2$_1$;

- $a_{21}b_{11}$ and $a_{21}b_{12}$ in RMCEX1$_2$ and RMCEX2$_2$;

- $a_{21}b_{21}$ and $a_{12}b_{22}$ in RMCEX1$_3$ and RMCEX2$_3$;

- $a_{11}b_{21}$ and $a_{11}b_{22}$ in RMCEX1$_4$ and RMCEX2$_4$;

- $a_{12}b_{11}$ and $a_{12}b_{12}$ in RMCEX1$_5$ and RMCEX2$_5$;

- $a_{11}b_{11}$ and $a_{11}b_{12}$ in RMCEX1$_6$ and RMCEX2$_6$.

Next and analogously, the coefficients of the first row of the third operator are loaded in registers RMCTP11 and RMCTP21 while those of the second row in registers RMCTP12 and RMCTP22. This done, multipliers MULTC1 and MULTC2 receive their inputs in the following order, enabling outputs from the involved registers, generating the operator coefficients for 3 q-bits given 2 basic 2×2 operator $C = [c_{ij}]$ for $1 \leq i, j \leq 2$. Table 5.2 shows the names of the registers that provide for the associated product computations for each of the 8 rows of the resulting quantum operator.

In each of the 8 clock cycles in which the results are being computed by MULTC1 and MULTC2 (each result is ready after 4 clock cycles once the operands are available). So, from the fifth to the thirteenth cycle, memory MSC is written with the pair of results (1 pair of results per address). Although the construction of an operator for 2 or more q-bits requires the execution of 4^q multiplications, where q is the number of q-bits of the operator under construction, there is a possibility that the matrix product between the operator and the column-vector, resulting from the tensor product of q-bits, is initiated before the completion of the operator construction. For this, it would be necessary for the q-bit tensor product to be ready and the last tensor product iteration between the previous operator for $q - 1$ q-bits and the basic quantum operator is current. Unit UCA would need to have two more multipliers for complex numbers for the tensor product.

5.4   Matrix Product

Recall that the matrix product is actually the quantum operation itself. Eventually preceded by the tensor products between operators and

Table 5.2: The used registers to provide the operand for the computation of the tensor product of 3 q-bits.

#Row	Register 1	Register 2	Register 3	Product 1	Product 2
8	RMC2	RMCTP12	RMCTP22	$a_{22}.b_{22}.c_{11}$	$a_{22}.b_{22}.c_{22}$
	RMC1	RMCTP12	RMCTP22	$a_{22}.b_{21}.c_{11}$	$a_{22}.b_{21}.c_{22}$
	RMCTP2	RMCTP12	RMCTP22	$a_{21}.b_{22}.c_{11}$	$a_{21}.b_{22}.c_{22}$
	RMCTP1	RMCTP12	RMCTP22	$a_{21}.b_{21}.c_{11}$	$a_{21}.b_{21}.c_{22}$
7	RMC2	RMCTP11	RMCTP21	$a_{22}.b_{22}.c_{11}$	$a_{22}.b_{22}.c_{12}$
	RMC1	RMCTP11	RMCTP21	$a_{22}.b_{21}.c_{11}$	$a_{22}.b_{21}.c_{12}$
	RMCTP2	RMCTP11	RMCTP21	$a_{21}.b_{22}.c_{11}$	$a_{21}.b_{22}.c_{12}$
	RMCTP1	RMCTP11	RMCTP21	$a_{21}.b_{21}.c_{11}$	$a_{21}.b_{21}.c_{12}$
6	RMCEX2$_1$	RMCTP12	RMCTP22	$a_{22}.b_{12}.c_{11}$	$a_{22}.b_{12}.c_{22}$
	RMCEX1$_1$	RMCTP12	RMCTP22	$a_{22}.b_{11}.c_{11}$	$a_{22}.b_{11}.c_{22}$
	RMCEX2$_2$	RMCTP12	RMCTP22	$a_{21}.b_{12}.c_{11}$	$a_{21}.b_{12}.c_{22}$
	RMCEX1$_2$	RMCTP12	RMCTP22	$a_{21}.b_{11}.c_{11}$	$a_{21}.b_{11}.c_{22}$
5	RMCEX2$_1$	RMCTP11	RMCTP21	$a_{22}.b_{12}.c_{11}$	$a_{22}.b_{12}.c_{12}$
	RMCEX1$_1$	RMCTP11	RMCTP21	$a_{22}.b_{11}.c_{11}$	$a_{22}.b_{11}.c_{12}$
	RMCEX2$_2$	RMCTP11	RMCTP21	$a_{21}.b_{12}.c_{11}$	$a_{21}.b_{12}.c_{12}$
	RMCEX1$_2$	RMCTP11	RMCTP21	$a_{21}.b_{11}.c_{11}$	$a_{21}.b_{11}.c_{12}$
4	RMCEX2$_3$	RMCTP12	RMCTP22	$a_{12}.b_{22}.c_{11}$	$a_{12}.b_{22}.c_{22}$
	RMCEX1$_3$	RMCTP12	RMCTP22	$a_{12}.b_{21}.c_{11}$	$a_{12}.b_{21}.c_{22}$
	RMCEX2$_4$	RMCTP12	RMCTP22	$a_{11}.b_{22}.c_{11}$	$a_{11}.b_{22}.c_{22}$
	RMCEX1$_4$	RMCTP12	RMCTP22	$a_{11}.b_{21}.c_{11}$	$a_{11}.b_{21}.c_{22}$
3	RMCEX2$_3$	RMCTP11	RMCTP21	$a_{12}.b_{22}.c_{11}$	$a_{12}.b_{22}.c_{12}$
	RMCEX1$_3$	RMCTP11	RMCTP21	$a_{12}.b_{21}.c_{11}$	$a_{12}.b_{21}.c_{12}$
	RMCEX2$_4$	RMCTP11	RMCTP21	$a_{11}.b_{22}.c_{11}$	$a_{11}.b_{22}.c_{12}$
	RMCEX1$_4$	RMCTP11	RMCTP21	$a_{11}.b_{21}.c_{11}$	$a_{11}.b_{21}.c_{12}$
2	RMCEX2$_5$	RMCTP12	RMCTP22	$a_{12}.b_{12}.c_{11}$	$a_{12}.b_{12}.c_{22}$
	RMCEX1$_5$	RMCTP12	RMCTP22	$a_{12}.b_{11}.c_{11}$	$a_{12}.b_{11}.c_{22}$
	RMCEX2$_6$	RMCTP12	RMCTP22	$a_{11}.b_{12}.c_{11}$	$a_{11}.b_{12}.c_{22}$
	RMCEX1$_6$	RMCTP12	RMCTP22	$a_{11}.b_{11}.c_{11}$	$a_{11}.b_{11}.c_{22}$
1	RMCEX2$_5$	RMCTP11	RMCTP21	of $a_{12}.b_{12}.c_{11}$	$a_{12}.b_{12}.c_{12}$
	RMCEX1$_5$	RMCTP11	RMCTP21	$a_{12}.b_{11}.c_{11}$	$a_{12}.b_{11}.c_{12}$
	RMCEX2$_6$	RMCTP11	RMCTP21	$a_{11}.b_{12}.c_{11}$	$a_{11}.b_{12}.c_{12}$
	RMCEX1$_6$	RMCTP11	RMCTP21	$a_{11}.b_{11}.c_{11}$	$a_{11}.b_{11}.c_{12}$

tensor products between q-bits, the matrix product is the last step in the set of computations to be performed by UCA. The matrix product results in the new coefficients of the set of q-bits, configuring the new quantum state.

As implemented, the matrix product is preceded by copying the q-bit coefficients in registers RMC1/RMC2, RMCPR1/RMCPT2, and the register series $RMCEX1_{1...n}/RMCEX2_{1...n}$, wherein $n = 2^{mq-1} - 2$. This initialization step is necessary so that the values of the coefficients before the operation are available for the repeated products with the coefficients of each row of the operator, releasing memory MQB to receive the new coefficients of the resulting product matrix. Furthermore, this prior register initialization has the advantage of having these coefficients locally available to unit UCA, avoiding repeated reads of memory MQB. The operation is executed according to the following steps:

1. The coefficients of the q-bit specified in the first quantum instruction descriptor are loaded into registers RMC1 and RMC2;

2. The coefficients of the q-bit specified in the second descriptor are loaded into registers RMCPT1 and RMCPT2;

3. The coefficients of the q-bit specified in the third descriptor and the following ones, if any, are copied into registers $RMCEX1_{1...n}$ and $RMCEX2_{1...n}$, according to the maximum number q-bits mq the co-processor can operate on simultaneously.

4. The basic operator coefficients in either memory MOP or MSC are read and inputted directly into multipliers MULTC1 and MULTC2 via multiplexers MUX31/MUX32 and MUX51/MUX52.

5. With the two available complex number multipliers, the matrix product can be performed for 2 coefficients at once (2 columns of an operator row, 2 rows of the column vector representing the q-bits). The pair of products is prepared simultaneously and added by adder SUMC1.

6. When executing the products related to the first 2 coefficients of each operator row, register RSC is reset so that adder SUMC2 passes on the same value contributed by SUMC1.

7. The result of SUMC2 is registered in RSC, which starts to function as an accumulator of sums of products, until the two last coefficients of the current row of the operator are dealt with. At this moment, register RSC has the final value of the coefficient to be stored at the appropriate addresses in memory MQB.

8. As each address of the MQB holds 2 coefficients (odd row and even row of the column-vector), when the matrix product regards an operator's odd row, the final result of the products registered in RSC is stored in RSP. When the result of the products of an even row of the operator is ready, the content of registers RSP and RSC are recorded in the appropriate address in memory MQB.

9. Once the multiplication of the last 2 coefficients of the operator is completed and the new coefficients of the vector-column are recorded in memory MQB, the co-processor becomes available to process the next request of the main processor, if any.

5.5 Chapter Considerations

In this chapter, we present the macro-architecture of the calculation unit UCA that performs all the multiplications and sums of complex numbers, means and end operations of used massively in quantum operations, especially when they involve entangled q-bits and/or operations that entangle q-bits. Unit UCA is designed in such a way as to make possible the parallelism of the multiplications between operator coefficients and q-bits. It is composed of two identical sets of components: multiplexers and registers augmented by the associated multipliers, adders, and registers. The ability to simultaneously process more coefficients is possible, with the replication of the aforementioned sets, which must be accompanied by some minimal changes in the number of inputs corresponding to the multiplexers.

The increase in the UCA ability in terms of simultaneous multiplications entails some changes in the handling of the provided results and the corresponding writing in the quantum bit memory as it will be explained in the next chapter. Therein, we describe the macro and micro-architecture of the control unit, which will hopefully complete the understanding of the full operation cycle of the quantum co-processor.

So, the next chapter is dedicated to the control unit of the quantum co-processor. It implements the dynamics necessary for the execution of quantum instructions.

Chapter 6

Control Unit Micro-architecture

In this chapter, we present the control unit of the processor. We first describe the micro-architecture of this unit and define the main control components required to implement the execution of the quantum micro-instructions. We also define the format of such micro-instructions. We then explain and exemplify the flow within the control unit during the execution of a given quantum instruction. After that, we do the same regarding the dynamics during the execution of the micro-instructions that interpret the quantum instructions. Subsequently, we present the tensor product controller, which controls the three main operations: tensor product of quantum bits; matrix product of quantum bits; tensor product of operators.

6.1 Introduction

The control unit UCO synchronizes the operation of the co-processor's remaining components mainly using a micro-program together with some auxiliary components. Based on the instructions prescribed by the main processor, unit UCO will orchestrate the other components of the macro-architecture as well as its internal components by means of planned micro-orders and some specific controllers. The micro-architecture of unit UCO is shown in Figure 6.1.

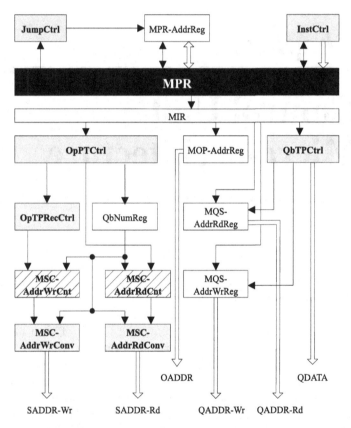

Figure 6.1: Micro-architecture of the control unit UCO.

Unit UCO includes one memory, two counters, six registers, five controllers, and two address converters. Their functions are detailed as follows:

1. JumpCtrl handles conditional and unconditional jumps within the micro-program;

2. InstCtrl is the instruction controller;

3. MPR is the read-only control memory wherein the micro-program resides;

4. MPR-AddrReg is the address register of the control memory;

5. MIR is the micro-instruction register;

6. OpTPCtrl controls the computation of the tensor product of quantum operator, starting from basic ones;

7. OpTPRecCtrl manages the recording of tensor product in MSC;

8. QbNumReg stores the quantity of current q-bits regarding the quantum operator in construction;

9. MOP-AddrReg is the address register of coefficients in memory MOP;

10. QbTPCtrl controls the computation of the tensor product of quantum bits as well as matrix product;

11. MSC-AddrRdCnt and MSC-AddrWrCnt are the address counters for reading and writing in memory MSC;

12. MQS-AddrRdReg and MQS-AddrWrReg are the address registers for reading and writing into memory MQS.

13. MSC-AddrRdConv and MSC-AddrWrConv are converters of reading and writing addresses of the tensor product coefficients in memory MSC, respectively.

Moreover, unit UCO handles SADDR-Rd and SADDR-Wr, which are the address buses for accessing memory MSC in read and write modes, respectively; OADDR, which is the address bus for reading form memory MOP; QADDR-Rd and QADDR-Wr, which are the address buses for accessing memory MQS in read and write modes, respectively; and QDATA, which is the main data bus for memory MQS.

The main processor sends a quantum operation, instruction by instruction. An instruction has a fixed size. A quantum operation is specified by as many instructions as there are q-bits to operate with.

The format of the quantum instruction informs three elements: the code of the quantum operator to be applied, which q-bit it must be applied to and whether it is the last instruction of the requested quantum operation. An instruction format is shown in Figure 6.2, wherein m denotes the number of available basic quantum operators, n the number of q-bits in the machine quantum state. Whenever the last bit of the instruction is 1, InstCtrl stops waiting for instructions. The basic operators used in this work are shown in Table 4.1.

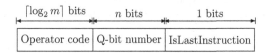

$\lceil \log_2 m \rceil$ bits	n bits	1 bits
Operator code	Q-bit number	IsLastInstruction

Figure 6.2: Format of a quantum instruction.

6.2 Instruction Flow

The set of instructions regarding a quantum operation as sent by the main processor is received and registered by component InstCtrl. The logic block of the instruction controller is shown in Figure 6.3. When activated input control signals 1 and 2 allow for the request of the next instruction of the quantum operation and reset the pointer to the instruction buffer, respectively. The output flags 3–5 depict the state of the instructions yet to be executed, if any. Output data signals 6–8 inform the current instruction being executed, the number of q-bits in the first instruction, and that of instructions in the current quantum operation, respectively. Output address signal 9 informs the last address to be read in memory MOP. Recall that input pins regarding data, address and control signals are shown in black, white and gray respectively, while the output pins regarding data, address and control signals are shown in hatched west, hatched east, and dotted respectively.

6.3 Micro-instruction Flow

The synchronization of the work performed by unit UCO to execute quantum operations is implemented using a micro-program, embedded into memory MPR. It is a sequence of micro-instructions, whose

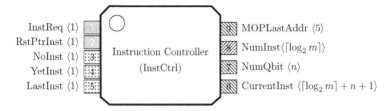

Figure 6.3: Logic block of component InstCtrl.

format is shown in Figure 6.4. It includes three main pieces of information:

1. Field *Micro-orders*, which is a set of orders used to trigger the components of the data-path, such as units UCA, UMS, read/write operations of differently designed memories, among others.

2. Field *Flags*, which is a set of status flags that are used in the micro-program instructions that require a conditional jump. There are two such instructions: the one that checks whether a specific flag is high and another one that checks whether the state of the flag is other than "high". In the current design, ten flags are used: MQBRdAddrZero and MQBWrAddrZero, indicating that the address to read/write memory MQB is zero; NoInst, denoting the list of empty instructions; PLusOp, indicating that there is more than one quantum operator in the current quantum operation; LastInst, informing the selection of the last instruction of the current quantum operation; EntangledBit, indicating that there is atleast one entangled q-bit in the set of specified q-bits; MSCRdAddrZero and MSCWrAddrZero, indicating that the address to read/write memory MSC is zero; TPOpComplete and TPQbComplete, indicating the termination of the tensor product of the operators listed in the instructions and that of the q-bits specified in the instructions.

3. The field address may represent four items: the address of a q-bit in memory MQB, an address in memory MOP, when it is necessary to access the coefficients of a basic quantum operator, an MPR address to jump to or a specific value to be loaded

in case it is necessary. The number of bits p reserved for this field is that which is sufficient to accommodate the largest of the four cases.

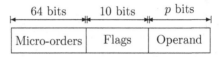

| Micro-orders | Flags | Operand |

Figure 6.4: Format of a micro-instruction.

The MPR address register is implemented as an up-counter with a preset option. The counter allows the sequencing of micro-instructions that do not entail deviation from the main flow of the program. Whenever a jump instruction occurs for which the jump condition is satisfied, the address informed in the micro-instruction is loaded in this counter. The correct update of the MPR address register is ensured by component JumpCtrl, whose self-explanatory implementation is shown in Figure 6.5.

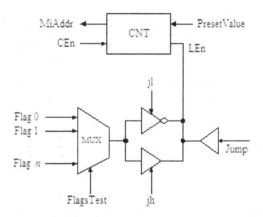

Figure 6.5: Micro-architecture of controller JumpCtrl.

The micro-program, stored in memory MPR, controls the operation of all components of the macro-architecture and also the auxiliary UCO data-path. This is done via the current micro-instruction available in register MIR. The micro-program main steps are sketched in Algorithm 1.

Algorithm 1 Main steps of the co-processor micro-program

1: __InitializeQbits__:
2: **Initialize** MQB address register with the address of the last q-bit
3: **repeat**
4:     **Write** 1.000 in ket $|0\rangle$ and 0.000 in ket $|1\rangle$ of the selected q-bit
5:     Increment MQB address register
6: **until** First q-bit is initialized
7: __WaitForInstruction__:
8: **repeat**
9:     **Read** half-duplex bus entry
10: **until** PROC presents instruction
11: __CollectInstruction__:
12: **repeat**
13:     **Log** instruction
14:     **Check** whether operator causes entanglement
15:     **Check** whether the specified q-bit is already tangled: $Entangled = 1$
16: **until** Receiving the last instruction: $T = 1$
17: __CheckIfTPRequired__:
18: **if** Operator causes entanglement and/or there is q-bit entanglement **then**
19:     **Jump to** *TensorProduct*
20: **else**
21:     **Jump to** *MatrixProduct*
22: **end if**
23: __TensorProduct__:
24: **Execute** tensor product between participating q-bits, creating a q-bit record
25: **if** Quantum operation does not cause entanglement **then**
26:     **repeat**
27:         *Execute* tensor product between participating operators
28:     **until** Finding the last instruction
29: **end if**
30: __productMatricial__:
31: **if** Operator causes entanglement and/or there is q-bit entanglement **then**
32:     **Get** operator from memory MSC
33:     **Multiply** the result by the q-bit register
34:     **repeat**
35:         **Multiply** each operator by the q-bit specified in the instruction
36:     **until** finding the last instruction
37: **end if**
38: **return** *waitForInstruction*

6.4 Tensor Product Controllers

Unit UCO includes specialized components to provide auxiliary control regarding the execution of tensor products between q-bits, as explained in Section 5.2, and tensor products between quantum operators, as described in Section 5.3, as well as the computation of matrix product as detailed in Section 5.4. The control regarding the tensor and matrix product of q-bits is described in Section 6.4.1 while the one regarding the tensor product of operators is explained in Section 6.4.2.

6.4.1 Controller of the Q-bit Tensor and Matrix Products

Component QbTPCtrl controls the read/write addresses for memory MQB, the control records of memory MQC as well as the multiplexers and registers of unit UCA so as it can produce the requested tensor product. The tensor product of q-bits, in addition to the proper mathematical operation, requires the treatment of the control records of the q-bits participating in the operation. Recall that such records, as defined in Figure 4.9, form a one-way linked list that includes all the information pertinent to the column-vector representation of the entangled q-bits.

Any operation that involves entangled q-bits will always handle the set of linked records starting from the first and read each one following the link order up to the last. The partial results of the tensor product of the set of entangled q-bits with the following q-bit, as defined in the associated records in MCQ, are made available at the outputs of multipliers MULTC1 and MULTC2 and placed on the data bus of memory MQB, preparing for a write operation. Then, component QbTPCtrl commands tensor multiplication of the coefficients of a set of q-bits indicated in the records with the q-bit indicated in the next record. Register MQS-AddrRdReg stores the address to be used in the read operation. The number of records required in MQB to store the coefficients of the column-vector that represents the entangled q-bits is 2^{q-1}, where q is the number of q-bits referenced in the associated records.

Component QbTPCtrl receives a trigger to initiate the tensor product, which when at the high level, makes QbTPCtrl take control of the co-processor components. The controller acts based on the following inputs:

1. The overall number of operators described in the current quantum operation;

2. The number of the q-bit defined in the first instruction of the current quantum operation;

3. The number of the q-bit of the current instruction;

4. The control record of the q-bit referenced by the current instruction;

5. The current micro-Instruction;

6. The control register where the new control data will be stored in MCQ;

7. The address of the q-bits where the data will be stored in MQB;

8. The address whose content is to be uploaded in the q-bit number register.

As soon as the controller receives the flag indicating that the requested tensor product computation has been completed, it enables the writing of the result into in memory MQS. Register MQS-AddrWrReg stores the address to be used in this write operation. It is noteworthy to point out that registers MQS-AddrRdReg and MQS-AddrWrReg are implemented as up-counters with a preset option, thus allowing for the sequential reads/writes during the initialization/measurement of q-bits, to setup/observe the machine quantum state. Moreover, the preset option allows for the registers to use a target address of an isolated read/write, as it can occur in a tensor product between q-bits and with a matrix product involving two or more q-bits.

6.4.2 Controller of Operator Tensor Product

Component OpPTCtrl controls the sequencing and synchronization of the read/write addresses of memory MSC as well as the multiplexers and registers of unit UCA so it can produce the requested composed operator from the basic ones available at memory MOP. Once triggered, this controller uses relative addresses, from the last to the first pair of coefficients that represent the quantum operator stored in the MSC memory. Based on a relative address handled by OpTPCtrl, converters MSC-AddrRdConv and MSC-AddrWrConv are responsible for determining the absolute address. The relative address from which the reading starts depends on the number of q-bits of the quantum operator currently in memory MSC. The number of addresses occupied in MSC by the current operator is given by $4^o q/2 = 2^{2oq-1}$, where oq is the number of q-bits that the operator handles. In this way, the relative addresses in the MSC starts from $4^o q - 1$ to 0. When constructing the operator for two q-bits, the two basic operators to be multiplied are read memory MOP, and the result of the tensor product is stored in memory MSC. During the construction of an operator for three or more q-bits, the operator read from memory MSC is multiplied tensorly by the operator provided by memory MOP. The result is written again in memory MSC. Reads and writes in MSC eventually occur simultaneously, but always at different addresses. During the construction of the operator for two or more q-bits, controller OpTPCtrl references the operator's coefficients following a predefined address sequence that starts at the first pair of coefficients in the first row and ends at the last pair of coefficients of the last row. Address converter MSC-AddrRdConv and MSC-AddrWrConv both compute the destination address in MSC using Equation 6.1:

$$A_{abs} = 2^{mq-oq} \left(A_{rel} - A_{rel} \bmod 2^{oq-1}\right) + A_{rel} \bmod 2^{oq-1}, \quad (6.1)$$

wherein A_{abs} and A_{rel} are the absolute and relative addresses, respectively; oq and mq denote the number of q-bits the operator being constructed will operate on and the maximum number of q-bits that the co-processor can handle simultaneously.

In order to illustrate organization and its usage of memory MSC during the construction of the quantum operator let us assume that the co-processor can operate on at most 4 q-bits simultaneously.

So, memory MSC has $2^{2 \times 4 - 1} = 128$ positions with rows $2^{4-1} = 8$ columns and $2^4 = 16$ rows. The memory positions are addressed as illustrated in Table 6.1, wherein, we show the organization of memory MSC as well as the addressable space for quantum operators of 2, 3 and 4 q-bits. So, in this example, we have $aq = 4$ with $oq = 2, 3$ or 4.

Table 6.1: Organization and absolute/relative addresses for memory MSC considering a quantum machine that can operate simultaneously on 4 q-bits atmost together with the positions occupied by an operator for 2 and 3 q-bits.

0	1	2	3	4	5	6	7
8	9	10	11	12	13	14	15
16	17	18	19	20	21	22	23
24	25	26	27	28	29	30	31
32	33	34	35	36	37	38	39
40	41	42	43	44	45	46	47
48	49	50	51	52	53	54	55
56	57	58	59	60	61	62	63
64	65	66	67	68	69	70	71
72	73	74	75	76	77	78	79
80	81	82	83	84	85	86	87
88	89	90	91	92	93	94	95
96	97	98	99	100	101	102	103
104	105	106	107	108	109	110	111
112	113	114	115	116	117	118	119
120	121	122	123	124	125	126	127

(a) $mq = oq = 4$ q-bits

0/0	1/1
8/2	9/3
16/4	17/5
24/6	25/7

(b) $oq = 2$ q-bits

0/0	1/1	2/3	3/3
8/4	9/5	10/6	11/7
16/8	17/9	18/10	19/11
24/12	25/13	26/14	27/15
32/16	33/17	34/18	35/19
40/20	41/21	42/22	43/23
48/24	49/25	50/26	51/27
56/28	57/29	58/30	59/31

(c) $oq = 3$ q-bits

It is up to OpTPCtrl to activate the appropriate control signals for multiplexers and registers of UCA so that the correct coefficients are placed at the inputs ports of multipliers MULTC1 and MULTC2. It also controls the enable signal regarding the write operation of the partial results of the operator tensor product in memory MSC. This signal stays activated during 4 clock cycles after the start of the tensor product, which is the latency of the multiplier of complex numbers. Controller OpTPCtrl provides the recording converters MSC-AddrRdConv and MSC-AddrWrConv with two signals that participate in determining the read/write address in memory MSC: the column number of the selected coefficient of the first operator and row number of the coefficient selected in the second operator.

Controller OpTPRecCtrl is responsible for recording the tensor product in memory MSC. It controls the sequencing of MSC addresses where the results of operator tensor products must be written. It uses the largest relative address of the next quantum operator, provided by a specific register, and of the relative address counter, MSC-AddrWrCnt. The counting enable signal is a micro-order provided by the current micro-instruction to activate this controller, starting the counting of clock cycles so that the recording of the results in memory MSC is enabled, taking into account the latency of the multiplier of complex numbers to produce the product. At the end of the countdown, the writing enable signal is activated, signaling that the writing phase of results has started. Once the relative address 0 is reached, the counter MSC-AddrWrCnt is reset.

Register QbNumReg is an up-counter that performs the function of a register. It provides the number of q-bits in the current operator. It is used by components MSC-AddrRdCnt, MSC-AddrWrCnt, MSC-AddrRdConv and MSC-AddrWrConv, so that the largest relative address of the operator under construction can be determined. At the end of each iteration of the sequence of tensor products between operators, register QbNumReg is incremented. When the operator construction starts, this component has its value initialized as 2.

6.5   Chapter Considerations

In this chapter, we present the design and operation of the Control Unit UCO, the unit of the co-processor architecture that manages all the remaining units as well as its own operation. Along with the Calculation Unit UCA, this unit forms the heart of the co-processor. It performs the expected control through micro-instructions and specialized components. On some occasions, the control is delegated to these components, but most of the time a simultaneous control is required.

The next chapter is dedicated to explaining the solution found to implement the quantum state measurement unit. The given solution is innovative and simple yet efficient.

Chapter 7

Quantum State Measurement

In this chapter, we propose a possible simple yet efficient implementation of the quantum state sampling unit. First, we present the idea behind the solution. Then, we describe and explain the microarchitecture of the measurement unit.

7.1 Introduction

The measurement unit UMS is responsible for readying the state of the quantum machine when requested by the main processor. The measurement of a quantum state results in the collapse of each q-bit of the co-processor into one of the states $|0\rangle$ or $|1\rangle$. No model for measuring quantum states was found in the literature regarding simulated quantum computing systems.

7.2 Measurement Unit

In this processor, we propose a simple way to implement the measurement task. Each of the 2^{nq} possible states for the co-processor

nq q-bits, with $nq \geq 1$, is associated with a given probability value. So, we idealized the measurement process based on the proportional selection model, also called *roulette*. It is most used in Genetic Algorithms during the selection phase of the individuals [17] to form the population of the next generation. Using the proportional selection model, each of the n individuals of the current pool has an associated score that is used to determine the probability of being selected for the next generation. In this work, we divide the probability interval [0, 1[, which is the range covered by the pseudo-random number generator, into n parts, each covering a sub-interval with no intersection. The extension of each sub-interval must be proportional to the score attributed to the respective individual [17]. Projecting this idea into the possible quantum states of the co-processor, one of which is identified as an individual, the probability associated with each quantum state would be used to define the extension of that quantum state. For example, a co-processor with 2 q-bits with the amplitudes presented in the second column of the table, shown in Figure 7.1(a) would have the associated sub-intervals defined according to the last column of the same table. The roulette wheel used in this case is shown in Figure 7.1(b).

The micro-architecture of the measurement unit UMS is shown in Figure 7.2. It includes a local control unit CUnit that synchronizes the different steps of the measurement operation. When signal *StartMs* is triggered, UMS receives the coefficients of each address in the MQB memory, denoted as OddReRd, OddImRd, EvenReRd and EvenImRd and using the functional unit RgCalc, starts performing the quantum state measurement. It first computes the higher limits of each sub-range for the given coefficients. Note the lower limit of the first sub-interval is always 0 and the following ones have the previous range's higher limit as their lower one. The 2^{nq} numbers relative to the sub-ranges, where nq is the number of q-bits of the co-processor, are then stored in the local memory LMem. It is implemented using a look-up table. Unit UMS computes the distribution of the sub-intervals and their extensions based on the result of the tensor product of all q-bits in MQB and the amplitudes obtained for each state. Then, it performs the comparison of latter values with the value provided by generator RNG, returning the selected quantum state that must be returned as the observed one.

State	Amplitude	Range
$\lvert 00 \rangle$	0.15	$[0.0, 0.15[$
$\lvert 01 \rangle$	0.20	$[0.15, 0.35[$
$\lvert 10 \rangle$	0.25	$[0.35, 0.6[$
$\lvert 11 \rangle$	0.4	$[0.6, 1.0[$

(a) quantum states and amplitudes

(b) Roulette wheel

Figure 7.1: Example of proportional selection for quantum state measurement.

It is noteworthy to emphasize, that for efficiency reasons, the pseudo-random number *rand*, generated by unit RNG of the macro-architecture, is compared in parallel with all the contents of all LMem positions using unit CMP, which is actually composed of 2^{nq} floats compare components. The match is declared for the state 2^i for which the comparison 1-bit result is set, yet that of state 2^{i-1} is reset. The quantum state thus selected is recorded in memory MQS via the setting of the first nq q-bits of MQB using either coefficient-tuple (1.0, 0.0, 0.0, 0.0) for the $\lvert 0 \rangle$ q-bits (0.0, 0.0, 1.0, 0.0) for the $\lvert 1 \rangle$ q-bits according to the binary representation of the observed state 2^i. Of course, the coefficients are configured according to the format of Figure 4.3. Recall, that all data is kept in IEEE754 standard floating number format. Moreover, the consequence of a measurement operation assumes a collapsed state for all the q-bits, and thus any eventual

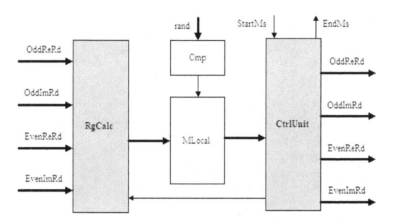

Figure 7.2: Micro-architecture of the measurement unit UMS.

entanglements must cease to exist. Thus, all the entries of memory MQC must be reinitialized, setting the content of each address *a* as $(a, a, 0)$, according to the MCQ word format of Figure 4.4. Once this process is completed signal *EndMs* is triggered to inform the end of the measurement operation.

In the proposed macro-architecture, component RNG is the pseudo-random number generator that provides a real value in the interval $[0, 1[$, based on a Linear Feedback Shift Register (LFSR) principle. The LFSR is a shift register whose input bit is the result of a linear function of its previous state. The only linear bit function available is binary XOR. So, it is an offset register whose input bit is driven by the XOR function of some bits of the register, allowing a random change in its value. The bit positions that affect the next state are called *taps*. The maximum period of the cycles over which the sequence is repeated is equal to $2n - 1$ for a LFSR register of *n* bits. Performing the function on the eight most significant bits of the mantissa and the 4 least significant bits of the exponent, as shown in Figure 7.3, guarantees that the numbers generated remain within the interval $[0, 1[$ [31, 7].

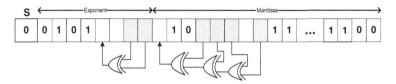

Figure 7.3: Configuration of the LFSR for the generation of random numbers in [0, 1].

It is noteworthy to point out that the size of local memory LMem in terms of bits can be defined as in Equation 7.1:

$$Size_{LMem} = 32 \times 2^{nq}, \tag{7.1}$$

7.3 Chapter Considerations

In this chapter, we present the design of the quantum state measurement, which is unit UMS, It performs the measurement of the quantum state using the roulette selection algorithm and pseudo-random number generator.

The next chapter presents simulation results regarding quantum instruction execution in the described quantum processor.

Chapter 8

Quantum Instruction Simulation

In this chapter, we give proof of the correct operation of the proposed processor. For this purpose, we present a simulation about every aspect of the quantum processor while executing a quantum instruction. We recall the main elements of a quantum instruction and their usage during the execution process. Then, we illustrate the execution of a quantum operation that manipulates a single quantum bit, showing the time diagram during its simulation. After that, we illustrate the execution of an operation on non-entangled quantum bits, showing the time diagram during its simulation. Then, follows the execution simulation of a quantum operation on entangled quantum bits. Subsequently, we illustrate the building of a high-dimensionality quantum operator from the predefined basic operator, showing the time diagram during its simulation. This is shown for two and three quantum bits. Then we, do the same for the tensor product of quantum bits, showing the time diagram during its simulation. Last but not the least, we illustrate via simulation of matrix product of entangled quantum bits and the measurement of the quantum state of the obtained result.

8.1 Introduction

The simulations presented in this chapter show the matrix product of a basic quantum operator of 1 q-bit and a q-bit and the tensor product of operators. It is arbitrated that the configuration of the quantum co-processor used in these simulations has 4 q-bits and capable of performing operations on 3 or 4 q-bits because it can properly demonstrate the operations and scalability of the design without resulting in the generation of very long time diagrams. Note that in several simulation snapshots presented in this Chapter, time diagrams with data-paths are repeatedly showing hexadecimal values 3F800000 and 00000000, corresponding to the complex coefficient $1 + 0i$ and $0 + 0i$ of quantum states with q-bits. Also, we assume that all memories, *i.e.*, MQB, MQC, MOP are initialized with their appropriate data contents as explained throughout Chapters 4–7.

8.2 Quantum Operations

Quantum operations are requested by the host processor PROC and can be performed on one or more q-bits, as specified in the instruction's descriptor. A quantum operation execution depends on three aspects:

1. The number of involved q-bits;

2. Whether one or more of the involved bits are entangled;

3. Whether the quantum operation causes entanglement.

Recall that each of the instructions is provided in signal *CurrentInstr*, which is one of the outputs of component InstrCtrl (see Figure 6.3). This signal will be seen in different time diagrams. An instruction is shown on the time diagram as three grouped values. For instance,: $\{02\}\{01\}\{1\}$. The first field ($\{02\}$ specifies the code for the quantum operation, the second ($\{01\}$), the number of the target q-bit of the quantum operation and, the third field ($\{01\}$) which defines whether the current instruction is the last one of the quantum operation. When this field has value 0, then the instruction is the last. Otherwise, when it has value 1, there are instructions still to be executed before the quantum operation is completed.

8.2.1 Operation on a Single Q-bit

Recall that a quantum operation involving a single 1 q-bit is executed as a matrix product between the specified basic quantum operator and the column vector representing the target q-bit. The coefficients of the quantum operator are available and read from memory MOP while the those regarding the state of the target q-bit are obtained from memory MQB. In the simulated example, the quantum NOT operation is applied on a q-bit still in the collapsed state $|0\rangle$, as defined in Equation 8.1:

$$NOT = \begin{bmatrix} 0 & 1 \\ 1 & 0 \end{bmatrix} \qquad NOT|0\rangle = \begin{bmatrix} 0 & 1 \\ 1 & 0 \end{bmatrix} \cdot \begin{bmatrix} 1 \\ 0 \end{bmatrix} = \begin{bmatrix} 0 \\ 1 \end{bmatrix} \qquad (8.1)$$

Figure 8.1 shows the time diagram for this operation. Signal InstReq commands the instruction controller InstCtrl to ready the next instruction. In the case of this example, there is only one instruction. Signal NoInst assumes a low level, as the current instruction is still in progress. Signal LastInst at a high level indicates that the current instruction is the last of the quantum operation. Signal CurrentInst is a record with three fields. In this example, the first field has a value of 2, which is the code of the NOT operator in this co-processor; the second field has value 2, which is the number of the target q-bit; and the last field is flag IsLastInst, whose value 0 indicates that it is the last instruction of the handled quantum operation. The coefficients of the target q-bit are read from memory MQB, which is in the state $|0\rangle$ (column vector with $1 + 0i$ in the first row and $0 + 0i$ in the second). From memory MOP, the coefficients of the operator NOT are obtained: in the first read cycle, the coefficients of the first row of the operator ($0 + 0i$ and $1 + 0i$) are obtained. In the second read cycle, the coefficients of the second row ($1 + 0i$ and $0 + 0i$) are fetched. First, the matrix product of the operator's first row is executed with the column vector: $[(0 + 0i).(1 + 0i) + (0 + 0i).(1 + 0i)]$, resulting in $0 + 0i$. Then, the matrix product of the second row of the operator is performed with the column vector: $[(1 + 0i).(1 + 0i) + (0 + 0i).(0 + 0i)]$, resulting in $1 + 0i$. The results are saved in memory MQB.

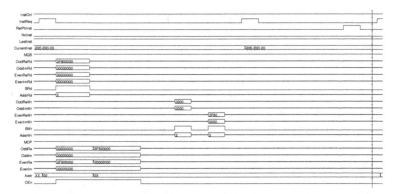

Figure 8.1: Matrix Product with NOT operator with one single target q-bit.

8.2.2 Operation on Non-entangled Q-bits

Quantum operation on non-entangled nq-bits with $n \geq 2$ is executed as n quantum operations on 1 q-bit each, performed sequentially, with an operation analogous to that described in Section 8.2.1. In the following example, the quantum operation NOT is applied on two non-entangled q-bits, which are collapsed into state $|0\rangle$. The NOT operator for two q-bits is shown in Equation 8.2:

$$NOT \otimes NOT = \begin{bmatrix} 0 & 1 \\ 1 & 0 \end{bmatrix} \otimes \begin{bmatrix} 0 & 1 \\ 1 & 0 \end{bmatrix} = \begin{bmatrix} 0 & 0 & 0 & 1 \\ 0 & 0 & 1 & 0 \\ 0 & 1 & 0 & 0 \\ 1 & 0 & 0 & 0 \end{bmatrix} \quad (8.2)$$

The column vector formed by the tensor product of two q-bits, each in state $|0\rangle$, is presented in Equation 8.3:

$$|0\rangle \otimes |0\rangle = \begin{bmatrix} 1 \\ 0 \end{bmatrix} \otimes \begin{bmatrix} 1 \\ 0 \end{bmatrix} = \begin{bmatrix} 1 \\ 0 \\ 0 \\ 0 \end{bmatrix} \quad (8.3)$$

The matrix-based representation of this operation is shown in Equation 8.4:

$$(NOT \otimes NOT)(|0\rangle \otimes |0\rangle) = \begin{bmatrix} 0 & 0 & 0 & 1 \\ 0 & 0 & 1 & 0 \\ 0 & 1 & 0 & 0 \\ 1 & 0 & 0 & 0 \end{bmatrix} \cdot \begin{bmatrix} 1 \\ 0 \\ 0 \\ 0 \end{bmatrix} = \begin{bmatrix} 0 \\ 0 \\ 0 \\ 1 \end{bmatrix} \quad (8.4)$$

Using the property of the tensor product shown in Equation 3.9 of Chapter 2, the NOT operation on two q-bits is performed in two steps, that is, two NOT operations (Equation 8.1), on each of the q-bits, separately. The tensor product between the column vectors resulting from each such NOT operation produces the column vector shown in Equation 8.3, in accordance with the aforementioned property of the tensor product. Figure 8.2 shows the time diagram of the two NOT operation execution on one q-bit since the instructions as requested. Figure 8.3 gives a amplified view of the time period related to the two operations. Figures 8.4 and 8.5 show, respectively, the time diagram of the first and second NOT operation.

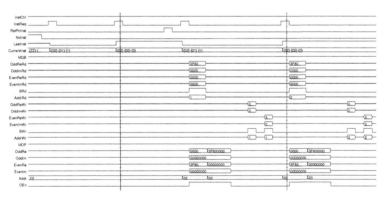

Figure 8.2: Matrix product of a NOT operation on two non-entangled q-bits.

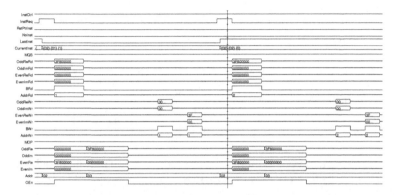

Figure 8.3: Amplified view of the two NOT operation on the q-bit: one after the other.

Each of the two descriptive instructions for the NOT operation, encoded in the signal CurrentInst, is presented as follows:

1. {02}{01}{1}: NOT operation on q-bit #1. This instruction is not the last of this operation.

2. {02}{02}{0}: NOT operation on q-bit #2. This instruction is the last of this operation.

Initially, signal NoInst is shown at a high level until an instruction is presented by the host processor. So, signal InstReq is triggered as many times as necessary until the last instruction in the current quantum operation is reached, checking whether there is at least one entangled q-bit involved in the current instruction. In the simulated example, it occurs twice. In the absence of an entangled q-bit, signal RstPtrInst is activated so that the instruction controller InstCtrl can supply the data of the first instruction the next time the signal InstReq command is executed. Each time the last instruction of the quantum operation is selected, signal LastInst shows a high level state. In the first activation of signal InstReq, the instruction pointer is initialized, as illustrated in the quantum operation on 1 q-bit in Section 8.2.1. As in the present example, we apply the quantum NOT operators on q-bits in the collapsed state $|0\rangle$. The calculations and results presented in the operation on one q-bit are repeated.

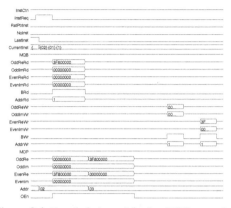

Figure 8.4: Detailed view of the first NOT operation.

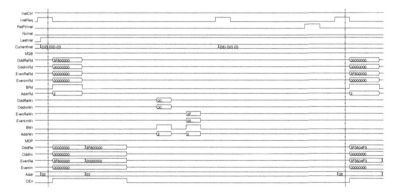

Figure 8.5: Detailed view of the second NOT operation.

8.2.3 Operation on Entangled Q-bits

When performing a quantum operation on two or more q-bits where at least one one of them is entangled with another and/or when the quantum operation causes entanglement, the construction of the quantum operator by means of a tensor product is required. In this section, examples of operator construction for two or more q-bits from basic quantum operators will be presented.

The coefficients of the basic quantum operators are read from memory MOP and stored in memory MSC when the operator is constructed for two q-bits. The construction of an operator for three or more q-bits requires the reading and writing of the coefficients in

MSC, eventually occurring simultaneously, but always at different addresses.

During the construction of the operator for two or more q-bits, the tensor product controller of quantum operators OpTPCntl references the operator coefficients following an address sequence that starts at the first pair of coefficients in the first row and ends at the last pair of coefficients of the last row of the operator matrix, as explained in Section 6.4.2 of Chapter 6. For instance, let us reconsider the quantum NOT operator for two q-bits, shown in Equation 8.4. Numbering the first cell of the first row as 0, and going through the cells from right to left and from top to bottom, allows obtaining a view of the organization of memory MSC with memory addresses as shown in Table 8.1 for the case of an operator of 2 q-bits, in Table 8.2 for the case of 3 q-bits and in Table 8.3 for 4 bits.

Table 8.1: Organization of memory MSC for two q-bits.

0	1
2	3
4	5
6	7

Table 8.2: Organization of memory MSC for three q-bits.

0	1	2	3
4	5	6	7
8	9	10	11
12	13	14	15
16	17	18	19
20	21	22	23
24	25	26	27
28	29	30	31

Table 8.3: Organization of memory MSC for four q-bits.

0	1	2	3	4	5	6	7
8	9	10	11	12	13	14	15
16	17	18	19	20	21	22	23
24	25	26	27	28	29	30	31
32	33	34	35	36	37	38	39
40	41	42	43	44	45	46	47
48	49	50	51	52	53	54	55
56	57	58	59	60	61	62	63
64	65	66	67	68	69	70	71
72	73	74	75	76	77	78	79
80	81	82	83	84	85	86	87
88	89	90	91	92	93	94	95
96	97	98	99	100	101	102	103
104	105	106	107	108	109	110	111
112	113	114	115	116	117	118	119
120	121	122	123	124	125	126	127

8.2.4 Building Quantum Operators

The case studies presented in the is section are regarding: *(i)* the construction of a quantum operator to operate with two q-bits when *(a)* memory MSC can accommodate at most operators to be used with three q-bits and *(b)* memory MSC can accommodate at most operators to operate with four q-bits; *(ii)* the construction of a quantum operator to operate with three q-bits when *(a)* memory MSC can accommodate at most operators to operate with four q-bits and *(b)* memory MSC can accommodate at most operators to operate with four q-bits.

8.2.4.1 Quantum Operator for Two Q-bits

In this example we assume that scratch memory MSC can hold at most an operator of 4 q-bits. So, with 128 addresses to accommodate the 256 coefficients of an operator for four q-bits, memory MSC in the present example will only have eight addresses occupied by the 16 coefficients of a NOT operator for two q-bits, built from two NOT operators for 1 q-bit. Initially, the coefficients of the first row are obtained from memory MOP, followed by the coefficients of the second row of the quantum operator NOT for 1 q-bit, being stored in the registers of the UCA calculation unit. In sequence, the coefficients of the second basic quantum operator are read from memory MOP and stored temporarily in unit UCA registers. The tensor product between these operators will generate the 8 coefficients to be stored in memory MSC. As it deals with reading and writing in different memories, the regressive addressing of MSC is not indispensable, which is necessary in the generation of operators for three or more q-bits. In this situation memory MSC is the source and destination of the data. In Table 6.1 in Chapter 6, the absolute addresses in bold are those occupied by the coefficients calculated for the operator that involves q-bits. The timing diagram for this operation is shown in Figure 8.6.

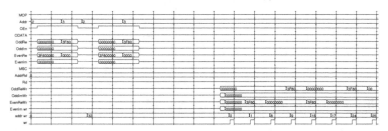

Figure 8.6: Building a NOT operator for 2 q-bits when the co-processor can handle quantum operations that involve at most 4 q-bits.

8.2.4.2 Quantum Operator for Three Q-bits

The construction of an operator for three q-bits with memory MSC sized for at most an operator of four q-bits must be preceded by the construction of an operator for two q-bits, as described in Section 8.2.4.1. Unlike the previous procedure, the data will come from both memories MSC and MOP. The operator for three q-bits has 64 coefficients, occupying only 32 addresses in memory MSC. Figure 8.7 shows the time diagram for this operation and Figures 8.8, 8.9 and 8.10 show the expanded views of the first overall time diagram of Figure 8.7. The addresses selected for simultaneous reading and writing regarding memory MSC are different and consistent with Table 6.1 in Chapter 6.

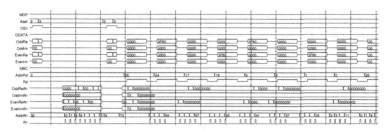

Figure 8.7: Building a NOT operator for 3 q-bits when the co-processor can handle quantum operations that involve at most 4 q-bits.

Figure 8.8: Details of the first part of the time diagram of Figure 8.7.

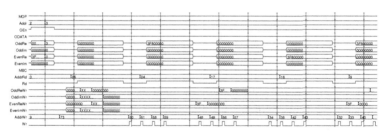

Figure 8.9: Details of the second part of the time diagram of Figure 8.7.

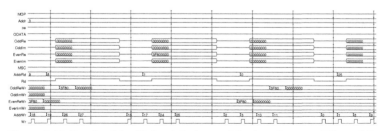

Figure 8.10: Details of the third and final part of the time diagram of Figure 8.7.

8.2.5 Tensor Product of Q-bits

When q-bits are entangled or are the target of an operation that causes entanglement, 2^n, where n is the number of q-bits participating in the operation, allocated memory positions are necessary for the coefficients of the column-vector representing the entangled q-bits, with two coefficients per memory address. The first MQB memory positions that were originally intended for bookkeeping the kets of each non-entangled q-bit, become useful for storing the initial pairs of coefficients that compose the entangled column-vector, with a complement of more memory positions whenever there are more than two involved q-bits. Recall that memory MQB has 2^{nq-1} positions, with two coefficients per memory address. For instance, consider $nq = 4$ and the tensor product on two q-bits ($n = 2$) that are in the collapsed state $|0\rangle$, as shown in Equation 8.5:

$$\begin{bmatrix} 1 \\ 0 \end{bmatrix} \otimes \begin{bmatrix} 1 \\ 0 \end{bmatrix} = \begin{bmatrix} 1 \\ 0 \\ 0 \\ 0 \end{bmatrix} \tag{8.5}$$

Table 8.4 shows the values of the coefficient pairs stored at each memory address in MQB before and after the tensor product. Note that in this example the same two memory positions are sufficient to store the four coefficients of the column vector that represents the result. Figure 8.11 shows the timing diagram regrading the aforemen-

Table 8.4: Contents of the first four addresses of memory MQB before and after performing the tensor product.

Address	Part	Original value	Computed value
0	Even	1.0000	1.0000
0	Odd	0.0000	0.0000
1	Even	1.0000	0.0000
1	Odd	0.0000	0.0000
2	Even/Odd	Unused	Unused
3	Even/Odd	Unused	Unused

tioned tensor product. Markers READ and WRITE are separators of the signal groups, associated with the read and write operations of memory MQB, respectively.

Figure 8.11: Memory MQB contents before and after a tensor product regarding the example of Equation 8.5 and Table 8.5.

8.3   Matrix Product of Entangled Q-bits

An operation on two or more entangled q-bits is performed by the matrix product of an operator matrix constructed at runtime and stored in memory MSC and a column vector stored in memory MQB at positions regarding the target q-bits. The example used for this simulation uses a constructed NOT operator, as defined in Equation 8.6:

$$\text{NOT} = \begin{bmatrix} 0 & 0 & 0 & 1 \\ 0 & 0 & 1 & 0 \\ 0 & 1 & 0 & 0 \\ 1 & 0 & 0 & 0 \end{bmatrix}, \tag{8.6}$$

Moreover, the hypothetical state of register R01 formed by two entangled q-bits (q-bit 0 and q-bit 1), is shown in Equation 8.7:

$$\text{R01} = \begin{bmatrix} 0.2237 \\ 0.4251 \\ 0.1039 \\ 0.8709 \end{bmatrix}. \tag{8.7}$$

The expected result of this operation is shown in Equation 8.8:

$$\text{NOT} \times \text{R01} = \begin{bmatrix} 0.8709 \\ 0.1039 \\ 0.4251 \\ 0.2237 \end{bmatrix} \tag{8.8}$$

In the time diagrams presented in this section, values are represented in hexadecimal form. In all involved computations, the floating point standard IEEE754 is used. Table 21 shows the correspondence of the float values used in this illustrative example according to this data format.

Table 8.5: Float values of the example in the used formatting standards.

Float value	Hexadecimal IEEE754
0.0000	0x00000000
0.2237	0x3E65119C
0.4251	0x3ED9A6B5
0.1039	0x3DD4C985
0.8709	0x3F5EF34D
1.0000	0x3F800000

The timing diagrams of Figures 8.12–8.15 illustrate the execution of the operation step by step. Therein, the markers READING MSC, MQB READING and UCA OUTPUT are used as separators of the signal groups, regarding those related to the buses and control signals related to memory MSC, memory MQB and unit UCA, respectively. Addresses 0 and 1 of memory MQB are read to obtain the coefficients of quantum register R01 and addresses 0/1, 2/3, 4/5 and 6/7 of memory MSC are read to obtain the coefficients of the operator.

Figure 8.12 shows the result of the multiplications required for the operator's first row. Figure 8.13 shows the result of the multiplications required for the operator's second row. Figure 8.14 shows the result of the multiplications required for the operator's third row. Figure 8.15 shows the result of the multiplications required for the operator's fourth row.

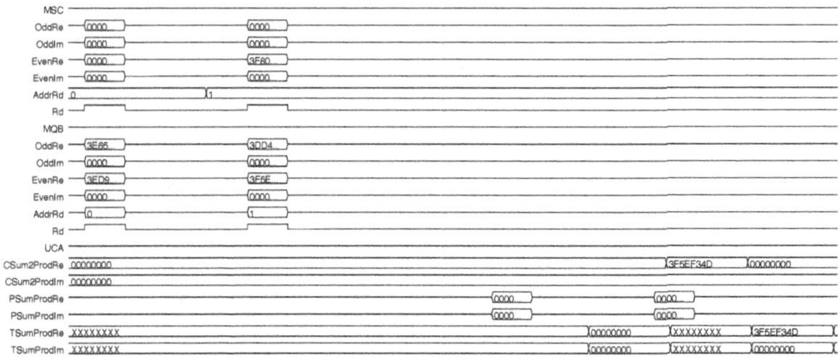

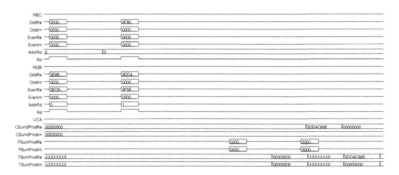

Figure 8.12: Matrix product result regarding the NOT operator's first row.

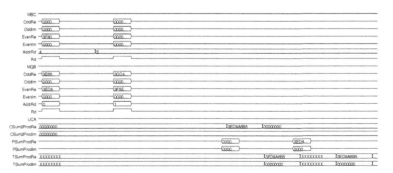

Figure 8.13: Matrix product result regarding the NOT operator's second row.

Figure 8.14: Matrix product result regarding the NOT operator's third row.

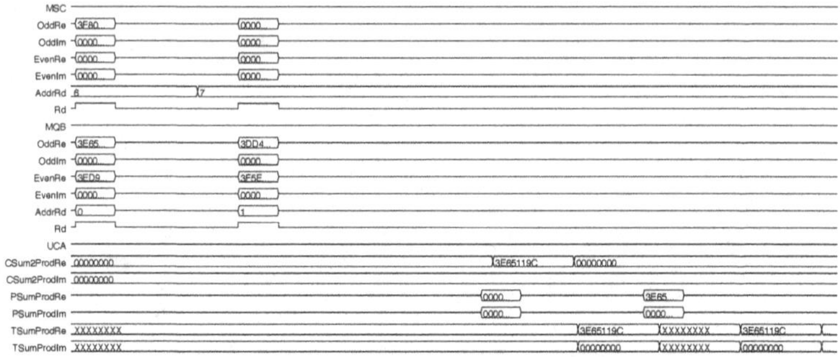

Figure 8.15: Matrix product result regarding the NOT operator's fourth row.

At the end of the quantum operation, register R01 is composed of the illustrated coefficients, as presented in Equation 8.8. These are written in memory MQB, as shown in the time diagram of Figure 8.16. Observe that the signals are all related to a writing cycle.

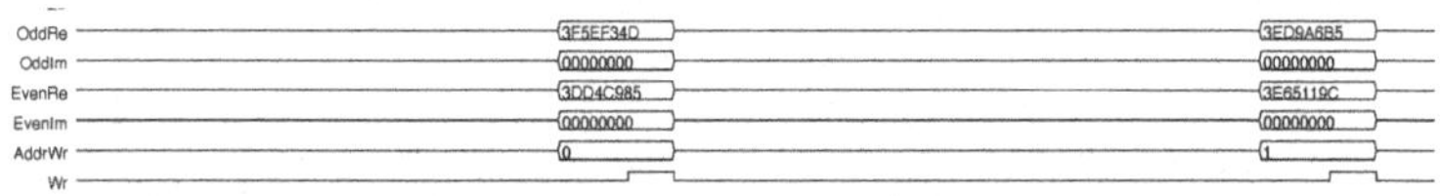

Figure 8.16: Matrix product result writing in memory MQB.

The four coefficients of register R01 once the NOT operation is concluded are associated with each of the four possible states expected as a result of a measurement of the quantum state. The sum of squares of the amplitudes totals 1, as proven in Equation 8.9:

$$0.87092^2 + 0.10392^2 + 0.42512^2 + 0.22372^2 = 1 \qquad (8.9)$$

Table 8.6 presents the interrelationship between the range of probability values and the quantum states shown in quantum register R01.

In this example, the quantum state measuring unit UMS receives the value 0.754657 from pseudo-random number generator (PRNG), which falls within the first range. So the observable state is therefore $|00\rangle$. This means that q-bits 0 and 1 in register R01 are in quantum state $|0\rangle$. This can be seen in Figure 8.17. The timing diagram shows the acquisition of the value provided by PRNG and the coefficients of the q-bits of the co-processor, followed by the writing of the result in memory MQB. In general, the measurement of the quantum state ends a cycle of operations of a quantum algorithm, resulting in a collapsed state for each q-bit of the co-processor, establishing a state comparable to that of a current general-purpose processor.

Table 8.6: Probabilities associated with each possible quantum state in register R01.

#range	Range interval	Probability (%)	Quantum state
1	[0.0000, 0.7585[	75.85	$|00\rangle$
2	[0.7585, 0.7693[	1.08	$|01\rangle$
3	[0.7693, 0.9500[	18.07	$|10\rangle$
4	[0.9500, 1.0000[	5,00	$|11\rangle$

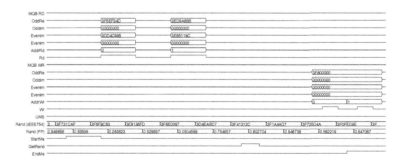

Figure 8.17: Matrix product result writing in memory MQB of the collapsed state of the q-bits after measurement.

8.4   Chapter Considerations

In this chapter, simulations involving tensor products of operators, tensor products of q-bits, matrix products of operators and representative column q-bit column vectors as well as measurement of quantum states are presented. It was shown how the coefficients of the quantum operator under construction are calculated: initially the coefficients of the basic quantum operators, which are stored in memory MOP are used then, the coefficients of the operator thus constructed, stored in memory MSC, and the coefficients of the next basic operator are used. The addresses read and written in memory MSC, for the same number of q-bits of the operator under construction, depend on the maximum number of q-bits that the co-processor can operate on simultaneously. Through the probability proportionate selection algorithm and pseudo-random number generator, it was possible to reproduce the randomness of the result respecting the amplitudes of each possible state. This concludes the proof via simulation that the proposed customizable quantum processor is functional and can be used to run quantum algorithms on general-purpose computers, allowing the evaluation of their effectiveness and performance.

The next chapter is dedicated to drawing some conclusions and pointing out some directions on how to improve the current state of the art regarding simulators/emulators of quantum computing.

Chapter 9

Final Book Remarks

In this book, basic elements of quantum computing are presented to help the reader in understanding the architecture and functionalities of the proposed co-processor design. Q-bit entanglement is a procedure used in quantum algorithms. It requires a massive performance of tensor products of operators and q-bits and matrix multiplications of operators and column-vectors representing the q-bits involved. It is important to emphasise that these are operations using complex numbers. The larger the number of q-bits participating in the operation, the higher the number of multiplications with complex numbers. This work shows an implementable approach in hardware with the possibility of scalability and parallelism explorations, which is essential to reduce the execution time of quantum operations.

9.1 Considerations

The proposed macro-architecture presents a number of components that could possibly be kept in a co-processor version with greater processing capacity. The calculation unit, responsible for executing the products and sums of complex numbers that are performed massively and extensively in the co-processor, is the unit of the co-processor that has the greatest impact on it overall performance. The modular characteristic of this unit allows for the proposed design to be expanded according to the desired level of parallelism. However, this implies an increase in the Control complexity, widening of the data path and, eventually, of buses.

The control unit is the component that manages all the remaining units present in the co-processor's data path, using control memory and auxiliary components. The control path automates some elementary tasks and contributes to simplify the microprogram.

The simulations validate the proper operation via some typical quantum operations and instructions that require matrix and tensor products, based on a scalable and parameterizable code. The simulations are run on a personal computer equipped with 4GB RAM memory and using the ModelSim program for Windows 7 operating system. In this case, it was possible to simulate a co-processor capable of operating up to 6 q-bits simultaneously.

9.2   Improvements

Several directions in terms of future work aiming at improving the co-processor design can be investigated, such that:

■ the co-processor can take advantage of more multipliers in the calculation unit, always a power of 2, in order to execute more products in the matrix and tensor products, simultaneously.

■ The pipeline organization of the architecture to streamline the quantum operation through the use of auxiliary complex number multipliers for the performance of the matrix product of the already calculated lines regarding the new operator under construction would reduce the wait time for the usage of all the required calculations.

■ The use of a scratch memory with more read and write ports, together with a larger number of multipliers of complex numbers, would accelerate the construction of the quantum operator.

■ The addition of a new auxiliary scratch memory to store operators constructed for more than one q-bit could save time in recalculating larger operators that are used recurrently.

References

[1] A full quantum eigensolver for quantum chemistry simulations. *Research—A Science Partner Journal*, 2020(1486935):1–11, 2020.

[2] Scott Aaronson. Quantum-Effect-Demonstrating Beef, 2011.

[3] Alexandre de Andrade Barbosa. Um simulador simbólico de circuitos quânticos. Mestrado, Universidade Federal de Campina Grande, 2007.

[4] Patrícia Silva Nascimento Barros. Reconhecimento de padrões quânticos aplicados à sequências de dna. Mestrado, Universidade Federal Rural de Pernambuco, 2011.

[5] André Berthiaume. Quantum Computation, 1996.

[6] Hamilton José Brumatto. Introdução à computação quântica. 2010.

[7] Rogerio Moraes Calazan, Nadia Nedjah, and Luiza de Macedo Mourelle. A massively parallel reconfigurable co-processor for computationally demanding particle swarm optimization. *LASCAS (2012)—International Symposium of IEEE Circuits and Systems in Latin America*, 2012.

[8] MacGregor Campbell. Quantum computer sold to high-profile client. *New Scientist Tech*, 2011.

[9] L.N.M. Carvalho, C. Lavor, and V.S. Motta. Caracterização matemática e visualização da esfera de bloch: Ferramentas para computação quântica. *Sociedade Brasileira de Matemática Aplicada e Computacional*, 2007.

[10] D-Wave. The D-Wave One system, 2012.

[11] K. De Raedt, K. Michielsen, H. De Raedt, B. Trieu, G. Arnold, M. Richter, Th. Lippert, H. Watanabe, and N. Ito. Massively parallel quantum computer simulator. *Computer Physics Communications*, 176(2):121–136, 2007.

[12] D. Deutsch. Quantum theory, the church-turing principle, and the universal quantum computer. *Proceedings of the Royal Society of London, Series A, Mathematical and Physical Sciences*, 400:97–117, 1985.

[13] M.W. Johnson et al. Quantum annealing with manufactured spins. *Nature*, (473):194–198, 2011.

[14] R.P. Feynman. Simulating physics with computers. *International Journal of Theoretical Physics*, 21(6):467–488, 1982.

[15] Nicolai Friis, Oliver Marty, Christine Maier, Cornelius Hempel, Petar Jurcevic Milan Holzäpfel, Martin B. Plenio, Marcus Huber, Christian Roos, Rainer Blatt, and Ben Lanyon. Observation of entangled states of a fully controlled 20-qubit system. *Physical Review X*, 8(2), 2018.

[16] X. Fu, L. Lao, K. Bertels, and C.G. Almudever. A control microarchitecture for fault-tolerant quantum computing. *Microprocessors and Microsystems*, 70:21–30, 2019.

[17] David Goldberg. *Genetic Algorithms in Search, Optimization and Machine Learning*. Reading, MA: Addison-Wesley Professional, 1989.

[18] Erico Guizzo. D-wave does not quantum compute. *IEEE Spectrum*, January 2010.

[19] Sara Hashemi, Mostafa Rahimi Azghadi, Ali Zakerolhosseini, and Keivan Navi. A novel fpga-programmable switch matrix

interconnection element in quantum-dot cellular automata. *International Journal of Electronics*, 102(4):703–724, 2015.

[20] I.G. Karafyllidis. Quantum computer simulator based on the circuit model of quantum computation. *IEEE Transactions on Circuits and Systems I: Regular Papers*, 52(8):1590–1596, 2005.

[21] A.U. Khalid, Z. Zilic, and K. Radecka. Fpga emulation of quantum circuits. In *IEEE International Conference on Computer Design: VLSI in Computers and Processors, 2004. ICCD 2004. Proceedings*, page 310–315, 2004.

[22] Mark LaPedus. 5/3nm wars begin, 2020.

[23] Jia Lee, Xin Huang, and Qing sheng Zhu. Decomposing fredkin gate into simple reversible elements with memory. *International Journal of Digital Content Technology and its Applications*, 4(5), 2010.

[24] Aércio Ferreira de Lima and Bernardo Júnior Lula. Circuitos quânticos: uma introdução. *WECIQ*, 2006.

[25] Felipe Lorenzen. Extensão dos be-ables de bell e competição atenuaççãoxamplificação. Master's thesis, Universidade de São Paulo, 2009.

[26] Guillermo Marcus. Floating point unit, 2004.

[27] A. MARON, Anderson Braga de Avila, Renata Hax Sander Reiser, and Maurício L. Pilla. Introduzindo uma abordagem para simulação quântica com baixa complexidade. In *X Brazilian Conference on Dynamic, Control and Applications*, page 1–4, 2011.

[28] Franklin de Lima Marquezino. A transformada de fourier quântica aproximada e sua simulação. Master's thesis, LNCC – Laboratório Nacional de Computação Científica, 2006.

[29] Kristel Michielsen. Juelicher supercomputer simuliert quantencomputer. March 2010.

[30] Matthias Möller and Cornelis Vuik. A conceptual framework for quantum accelerated automated design optimization. *Microprocessors and Microsystems*, 66:67–71, 2019.

[31] Nadia Nedjah and Luiza de Macedo Mourelle. An efficient problem-independent hardware implementation of genetic algorithms. *Neurocomputing*, 71(1):88–94, 2007.

[32] M.A. Nielsen and I.L. Chuang. *Quantum Computation and Quantum Information*. Cambridge, UK: Cambridge University Press, 2000.

[33] Eesa Nikahd, Mahboobeh Houshmand, Morteza Saheb Zamani, and Mehdi Sedighi. One-way quantum computer simulation. *Microprocessors and Microsystems*, 39(3):210–222, 2015.

[34] Ivan S. Oliveira and Roberto S. Sarthou. Computação quântica e informação quântica. *V Escola do CBPF*, 2004.

[35] Nigini Abilio Oliveira. A utilização do algoritmo quântico de busca em problemas da teoria da informação. Master's thesis, Universidade Federal de Campina Grande, 2007.

[36] Bernhard Ömer. A procedural formalism for quantum computing. Master's thesis, Technical University of Vienna, 1998.

[37] Bernhard Ömer. Quantum programming in qcl. Master's thesis, Technical University of Vienna, 2000.

[38] Bernhard Ömer. Classical concepts in quantum programming. *International Journal of Theoretical Physics*, 44(7):943–955, 2005.

[39] Bernhard Ömer. *Structured Quantum Programming*. PhD thesis, Vienna University of Technology, 2009.

[40] Renato Portugal, Carlos Magno Martins Cosme, and Demerson Nunes Gonçalves. *Algoritmos Quânticos*. 2006.

[41] Renato Portugal, Carlile Lavor, Luiz Mariano Carvalho, and Nelson Maculan. *Introdução a computaçãao quantica, Notas de Matematica Aplicada, Vol. 8*. SBMAC, São Carlos, 2004.

[42] Eleanor Rieffel and Wolfgang Polak. An introduction to quantum computing for non-physicists. *ACM Comput. Surveys*, 32:300–335, 2000.

[43] Gustavo Rigolin. Emaranhamento quântico. *Revista Phisicae*, 7, 2008.

[44] David Schor. Tsmc starts 5-nanometer risk production. *WikiChip Fuse*, 2019.

[45] V.V. Shende, S.S. Bullock, and I.L. Markov. Synthesis of quantum-logic circuits. *IEEE Transactions on Computer-Aided Design of Integrated Circuits and Systems*, 25(6):1000–1010, 2006.

[46] Fernando Luís Semião da Silva. Processamento de informação quântica usando um sistema de íons aprisionados e cavidades. Master's thesis, Universidade Estadual de Campinas, 2002.

[47] Ronaldo Thibes. Álgebra linear e mecânica quântica. *V Bienal da SBM*, 2010.

[48] J.P.G. van Dijk, E. Charbon, and F. Sebastiano. The electronic interface for quantum processors. *Microprocessors and Microsystems*, 66:90–101, 2019.

[49] André Luís Vignatti, Francisco Summa Neto, and Luiz Fernando Bittencourt. Uma introdução à computação quântica, 2004.

[50] Juliana Kaizer Vizzotto and Antônio Carlos da Rocha Costa. Linguagens de programação quântica - um apanhado geral. *WECIQ*, 2006.

[51] Mário Sansuke Maranhão Watanabe. O algoritmo polinomial de shor para fatoração em um computador quântico. Master's thesis, Universidade Federal de Pernambuco, 2003.

[52] Robert Wille, Eleonora Schönborn, Mathias Soeken, and Rolf Drechsler. Syrec: A hardware description language for the specification and synthesis of reversible circuits. *Integration*, 53:39–53, 2016.

[53] W.K. Wooters and W.H. Zurek. A single quantum cannot be cloned. *Nature*, 1982.

[54] X.C. Wu, S. Di, M.E. Dasgupta, F. Cappello, H. Finkel, Y.A. Alexeev, and F.T. Chong. Ifull-state quantum circuit simulation by using data compression. In *Proceedings of the International Conference for High Performance Computing, Networking, Storage and Analysis*, number 80, page 1–24, 2019.

[55] Guowu Yang, William N.N. Hung, Xiaoyu Song, and Marek A. Perkowski. Exact synthesis of three-qubit quantum circuits from non-binary quantum gates. *International Journal of Electronics*, 97(4):475–489, 2010.

[56] Noson S. Yanofsky. An introduction to quantum computing. *arXiv:0708.0261*, pages 1–33, 2007.

Index

Printed in the United States
by Baker & Taylor Publisher Services